General R

General Relativity

I. R. KENYON
School of Physics and Space Research
Birmingham University

Oxford New York Tokyo
OXFORD UNIVERSITY PRESS

Oxford University Press, Walton Street, Oxford OX2 6DP
Oxford New York Toronto
Delhi Bombay Calcutta Madras Karachi
Petaling Jaya Singapore Hong Kong Tokyo
Nairobi Dar es Salaam Cape Town
Melbourne Auckland
and associated companies in
Berlin Ibadan

Oxford is a trade mark of Oxford University Press

Published in the United States
by Oxford University Press, New York

First published 1990
Reprinted 1991

British Library Cataloguing in Publication Data
Kenyon, I. R. (Ian R.)
General relativity.
1. Physics. General theory of relativity
I. Title
530.11
ISBN 0-19-851995-8
ISBN 0-19-851996-6 (pbk)

Library of Congress Cataloging in Publication Data
Kenyon, I. R.
General relativity / I. R. Kenyon.
Includes bibliographical references.
1. General relativity (Physics) I. Title.
QC173.6.K46 1990
530.1'1—dc20 89-49584 CIP

ISBN 0-19-851995-8
ISBN 0-19-851996-6 (pbk)

Set by
Asco Trade Typesetting Ltd, Hong Kong

Printed and bound by
Biddles Ltd, Guildford and King's Lynn

To Valerie

Acknowledgements

I would like to thank colleagues who generously agreed to read sections of the draft version of the text; their comments helped to improve the text substantially. I am particularly indebted to Professor John Dowell, Dr Michael Church, Dr Peter West, Dr Goronwy Jones and Dr Raymund Jones.

Professors V. Rubin, J. H. Taylor, R. Isaacman, A. Lange, E. Amaldi, I. I. Shapiro, K. Kuroda and J. Surdej were kind enough to permit the reproduction of diagrams from their publications. My thanks also go to the Societa Italiana di Fisica (*Il Nuovo Cimento*), the American Physical Society (*Physical Review Letters* and *Review of Modern Physics*), *Nature*, Cambridge University Press, and the Reidel Publishing Company for permission to reproduce material. Miss S. Ellis word processed the text and the many corrections with great accuracy; I am extremely grateful for her help. My warm thanks also go to Mrs M. Baggott who produced the excellent originals for the illustrations.

Preface

Einstein's development of the general theory of relativity was roughly contemporary with the genesis of quantum mechanics. Yet of these two outstanding developments in physics the second receives full and adequate coverage in current undergraduate courses in science while the first is little mentioned. The difference in emphasis comes about partly because the applications of general relativity are remote from day-to-day life and science, and partly because of the mathematical content of the subject. Despite these inhibiting factors it seems to the author that science undergraduates other than specialist mathematicians ought to have some exposure to this other triumph of physics, and it is the aim of this volume to help make this possible.

Experiments using modern technology have achieved astonishing results relevant to the study of general relativity; for example, the evidence for gravitational radiation from a binary pulsar 15 000 light years distant from the earth, PSR 1913 + 16. Therefore it is important to give full prominence to the developing experimental aspects of the subject.

Chapter 1 is an introduction and outline with a review of the relevant results and notation of special relativity. In the next three chapters the equivalence principle and the ideas of space–curvature are introduced; this culminates in a simplistic presentation of the Schwarzschild metric. At this point the reader has to hand the description of space—time around stars and simple black holes. Three chapters giving a physically oriented introduction to the formalism of Riemannian geometry then follow. After this there is a chapter on tests of general relativity inside and beyond the solar system. In Chapter 9 the strange properties of black holes are discussed, and the evidence for their observation. Next there is an account of gravitational radiation and the experiments designed to detect it. Chapter 11 describes the interplay of the general theory of relativity and cosmology. The final chapter assesses attempts to quantize the gravitational field.

Appendices are used to present reference material. This includes details on the properties of metric components, connections and curvature tensors, and a table of physical constants. A set of problems with complete worked solutions is given in Appendix I. To close there is an annoted bibliography and a set of references to original papers.

For a course with reduced mathematical content Chapters 5–7 could either be skipped or used sparingly.

Birmingham
September, 1989

I.R.K.

$2GM/c^2$ of the centre is inevitably trapped by the intense gravitational field: space–time is so warped that not even light can escape.

GR also predicts the existence of gravitational waves which travel with the velocity of light. When they cross a region of space–time it is space–time itself that vibrates. Space–time is remarkably rigid, so that the amplitude of the strain is only 10^{-18} when the pulse from a supernova collapse at the centre of the Galaxy reaches the Earth. Detectors are only now approaching the sensitivity required to detect such tiny oscillations. However, there is indirect evidence that the orbit of a binary pair, one member of which is the pulsar PSR 1913+16, is collapsing at exactly the rate expected due to the loss of energy in gravitational radiation. In cosmology GR has been crucial in providing a framework for discussing the structure of the Universe, for example the way that the energy content of the Universe, its expansion rate, and the curvature of space–time are related. It is also worth mentioning how Mach's principle is explained in this context. One observation which illustrates this principle is that a Foucault pendulum at the North Pole swings in a plane fixed relative to the frame of the distant galaxies; however, an earth-bound observer sees the plane of swing rotate through 360° every 24 hours. Somehow the matter in the Universe seems to determine the existence of an inertial frame in which Newton's laws of motion are valid. This idea has been reinforced by observations made on the cosmic background radiation (CBR). The CBR is black-body radiation at 2.74 K that bathes the Earth's orbit and presumably permeates the whole Universe. There is an anisotropy in the radiation which is explained as a Doppler shift due to the Earth's motion relative to the frame of the distant galaxies; the radiation would appear isotropic in that frame. As far as the CBR is concerned the distant galaxies provide a preferred frame, something that is anathema to SR or classical mechanics. In modern cosmology a preferred frame emerges naturally (the comoving frame) for universes whose behaviour is consistent with GR.

It is interesting to contrast gravitation with the other long-range force of nature—electromagnetism. The long range of each force is attested in one case by the operation of the solar system and in the other by the presence of the Van Allen belts of charged particles trapped in the Earth's magnetic field. Of the two the electromagnetic force is intrinsically far stronger: for instance the electric repulsion of two protons is 10^{36} times stronger than their gravitational attraction. It seems surprising that the gravitational force should dominate in the Universe. However, all matter (and anti-matter) exerts a gravitational attraction, while positive and negative electric charges exert opposite effects. In an atom the electrons effectively screen the nuclear charge so that a feeble van der Waals force penetrates outside a neutral atom, and this falls off with distance as r^{-6}. The very fact that the gravitational attraction dominates tells us that matter on the large scale must be net electrically neutral to very high precision.

1.1 Outline of contents

Physical concepts are made the basis for this text on the development and achievements of GR. The reader should realize that Einstein did not give a rigorous mathematical proof of GR and neither has anyone else. His approach involved physical insight and the use of simple arguments. Einstein learnt the mathematical techniques he needed from his friend Marcel Grossman over a number of years when they both lived in Zurich and regarded the techniques primarily as useful tools. Some differential geometry and tensor analysis is needed in order to appreciate the full flavour of the theory. In the text below an attempt is made to restrict the discussion of these techniques to those features essential to the understanding of GR.

Einstein's starting point was the equivalence principle and this concept is made the subject of Chapter 2. Briefly, the principle states that physics appears the same to any observer in free fall whatever the magnitude of the gravitational field. In the observer's frame the physics is described by SR. When the equivalence principle is applied to photons they are predicted to undergo a gravitational red shift when they leave a star, whilst a deflection is expected for light passing near to a massive body. These effects are interpreted as arising from the distortion of space–time in the presence of matter. In Chapter 3 examples of the geometry of curved spaces are discussed, beginning with two-dimensional surfaces. Techniques for handling curved space–time are discussed in Chapter 4, which ends with a presentation of the Schwarzschild solution for space–time outside a spherically symmetric mass distribution. This is a key result because it covers the cases of most interest: the solar system effects and simple black holes.

In developing the formal mathematical structure of GR Einstein made use of the existing theory of curved spaces discovered by Gauss, Riemann, and others in the nineteenth century. This employs the techniques of tensor analysis which are introduced in Chapter 5. The general theory is then developed from this basis in Chapters 6 and 7. A major feature is that real space–time is Riemannian, that is to say it is curved yet locally looks like the flat space–time of special relativity much as a portion of a spherical surface of linear extent much smaller than the radius looks flat. In each local region of space–time a frame in free fall can be selected in which SR is valid. Riemannian space–time is a patchwork of such regions. One of Einstein's insights was to infer that because mass generalizes to mass/energy in SR it must be this mass/energy that causes the curvature of space–time. The most compact description of mass/energy is provided by the stress-energy tensor. Einstein proposed a simple relationship

$$\text{curvature tensor} = \left(\frac{8\pi G}{c^4}\right) \text{stress-energy tensor},$$

but it took him a number of years to identify the precise form of the curvature tensor involved. In the limit of slowly moving bodies and weak gravitational fields this equation will be shown to reduce to Newton's law of gravitation.

Further tests of GR are described in Chapter 8: one involves the anomalous precession in the orbits of Mercury and of the binary pulsar 1913+16, another concerns delays of planetary radar reflections passing near the Sun, and finally there is the gravitational lensing of a quasar image by galactic matter. Chapter 9 is devoted to black holes. We follow a probe into the interior of a black hole and find that space and time interchange some of their characteristics under the intense gravitational field present there. The formation of black holes and evidence for their existence are also covered.

In Chapter 10 we discuss the prediction of the existence of gravitational waves, which are transverse and travel at the speed of light in empty space. Gravitational waves are quite simply the vibration of space–time itself. The gravitational constant G determines the strength of the coupling of these waves to space–time; its value is small and we shall see that this implies that space–time is very stiff. Modern detectors which should be capable of detecting the gravitational waves from a supernova collapse in our Galaxy are described. One variety uses a freely suspended massive bar cooled to 4 K and another variety uses a large Michelson interferometer monitored with laser beams. The observation of slowing down of the orbital rotation of the binary pulsar 1913+16 is also discussed. There is no classical explanation for the slowing down; however, the rate of energy loss is exactly that expected via gravitational radiation.

In chapter 11 the impact of GR on cosmology is described. Robertson and Walker proposed a model for the Universe having uniform curvature; this framework works very well and leads to the locally preferred comoving frames noted above. In the present epoch the systematic red shift of galactic spectra indicates that the Universe is expanding. Applying Einstein's equation to the Robertson–Walker model of space–time in which the Universe is pictured as filled with an ideal fluid permits us to follow the dynamics of this model Universe. For example, we can relate the density of matter in the Universe to the rate of its expansion. Only general conclusions can be drawn at present because not enough is known about the parameters involved in this model; whether the Universe will expand forever or whether it will finally contract is not yet clear.

Chapter 12 contains a discussion of the possibility of quantizing the gravitational force in the same way that the electromagnetic and other forces are quantized. Special difficulties arise because the structure of space–time becomes quantized. The problems and possible schemes for their resolution mark the end of the story so far.

1.2 Summary of results in special relativity

The two basic postulates of SR are as follows: firstly the laws of physics take the same form in all inertial frames, i.e. in all frames which are moving with constant velocity with respect to the frame of the distant galaxies; secondly, the velocity of light is a constant c. A point in space–time (which we shall call an event) has coordinates x, y, z with respect to a rectangular Cartesian set of space axes at time t. These four space–time coordinates form a four-vector with components

$$x^0 = ct, \qquad x^1 = x, \qquad x^2 = y, \qquad x^3 = z.$$

We also write $\mathbf{r}$ for the spatial three-vector.

In another inertial frame moving with relative velocity $v = \beta c$ parallel to the x axis (x' axis) the new (primed) coordinates of the same event are given by the Lorentz transformation

$$\left.\begin{aligned} x^{0\prime} &= \gamma(x^0 - x^1\beta) \\ x^{1\prime} &= \gamma(x^1 - x^0\beta) \\ x^{2\prime} &= x^2 \qquad\qquad x^{3\prime} = x^3 \end{aligned}\right\} \tag{1.1}$$

where $\gamma = 1/(1-\beta^2)^{1/2}, 0 \leqslant \beta < 1$ and $\gamma \geqslant 1$. The interval between two events P_1 and P_2 with coordinate separation $(\Delta x^0, \Delta x^1, \Delta x^2, \Delta x^3)$ is defined to be

$$\begin{aligned} \Delta s^2 &= (\Delta x^0)^2 - (\Delta x^1)^2 - (\Delta x^2)^2 - (\Delta x^3)^2 \\ &= (\Delta x^0)^2 - \Delta r^2 \end{aligned}$$

and is easily shown to be invariant under Lorentz transformations. If a clock travels from P_1 to P_2 then the time interval it measures between these events is called the *proper time* $\Delta\tau$, with $c\Delta\tau = \Delta s$. Note also that the components of $\Delta\mathbf{r}$ transverse to the velocity vector are not affected. There are other important four-vectors, for example the four-momentum with components

$$p^0 = E/c, \qquad p^1 = p_x, \qquad p^2 = p_y, \qquad p^3 = p_z,$$

where E is the total energy and $\mathbf{p}$ is the relativistic linear momentum. The Lorentz transformation takes the same form for all four-vectors: in the case of the four-momentum we can take the formulae (1.1) given above for the coordinates and replace x by p throughout. Each four-vector has an invariant, which in the case of the space–time interval P_1P_2 is Δs^2. The invariant of any other four-vector is formed in the same way by taking the difference of the squares of its time and space components. Hence the four-momentum vector has an invariant

$$(p^0)^2 - (\mathbf{p})^2 = E^2/c^2 - \mathbf{p}^2 = m^2c^2,$$

where m is the rest mass of the body involved. Like Δs^2 this is invariant under any Lorentz transformation. The four-momentum is related to the four-velocity whose components are

$$v^0 = c\gamma, \qquad v^1 = v_x\gamma, \qquad v^2 = v_y\gamma, \qquad v^3 = v_z\gamma.$$

Here, v_x, v_y, and v_z are the standard components of velocity, meaning that in time dt the distance travelled along the x direction will be $v_x\,dt$. Also $\gamma = 1/(1 - \beta^2)^{1/2}$, where $\beta c = (v_x^2 + v_y^2 + v_z^2)^{1/2}$. The invariant of the velocity four-vector is c^2. For a body of mass m the relation between the momentum and velocity four-vectors is

$$p = mv,$$

as in the classical case. There is also a four-vector force F defined by using the relativistic equivalent of Newton's second law of motion

$$F = \frac{dp}{d\tau}, \tag{1.2}$$

where we note that the differentiation is with respect to the proper time τ which is a Lorentz scalar. Then

$$F^0 = \frac{dE}{c\,d\tau} = \gamma\frac{dE}{c\,dt}$$

$$F^1 = \frac{dp_x}{d\tau} = \frac{\gamma\,dp_x}{dt} = \gamma f_x, \text{ and so on,}$$

where f_x, f_y, and f_z are the components of the classical three-dimensional force. Thus the four-force is

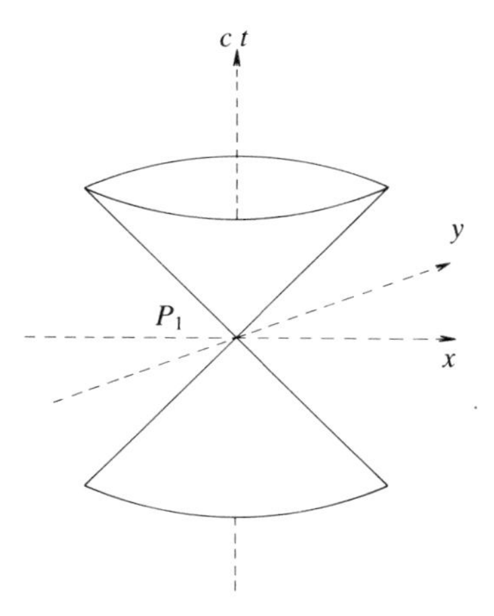

Fig. 1.1 The light cone through a point P_1 (one space dimension is suppressed).

$$F = \gamma\left(\frac{\mathrm{d}E}{c\,\mathrm{d}t}, f_x, f_y, f_z\right).$$

The interval Δs^2 can be positive, negative, or zero. If Δs^2 is zero, so that $c\Delta t$ exactly equals Δr, then a light ray can travel from event P_1 to event P_2 and the interval $P_1 P_2$ is called light-like. When $c\Delta t$ exceeds Δr, the interval is positive and P_2 can be reached from P_1 by travelling at a velocity less than the speed of light; the interval is called time-like because we can choose an inertial frame such that $\mathbf{r}_1 = \mathbf{r}_2$ leaving only a time separation between the two events.

Finally if the interval Δs^2 is negative $c\Delta t$ is less than Δr. Then no information can pass from P_1 to P_2 because it would need to travel faster than light itself. Such a separation is called space-like, and in this case it is always possible to find an inertial frame such that $t_1 = t_2$ with the separation being a spatial distance. In this case whatever happens at P_1 can have no influence on P_2, and vice versa. Figure 1.1 illustrates these different types of interval; time is the upward axis and the others are spatial (one spatial direction is suppressed). The cone is drawn with

$$ct = (x^2 + y^2)^{1/2}$$

and marks out all the possible light paths through P_1; it is called the light cone at P_1. Inside the forward light cone ($t > 0$) lie all the space–time events with time-like separation from P_1 which can be influenced by what happens at P_1. Outside the light cone are events at space-like separations from P_1, and these cannot ever communicate with P_1. Inside the backward light cone ($t < 0$) are the events that influence what happens at P_1.

2
The equivalence principle

A number of experiences familiar to modern man indicate that there is a close resemblance between the gravitational force and the effects of acceleration. High-speed centrifuges generate large inertial forces which are used to separate materials from liquid suspensions that would sediment only slowly, if at all, under gravity. Pilots of jet aircraft making tight turns feel forces that are labelled 'g forces', and in the realm of space exploration proposals exist to build giant wheels that would rotate to provide an artificial gravity. These are all examples of centrifugal acceleration. Linear acceleration is less readily sustained and has fewer familiar applications. However, the parallel between a linear acceleration and gravitation is conceptually simpler. Einstein realized that this parallel is a principle of nature, the *equivalence principle*, which states that a region of uniform gravitational field and a uniformly accelerating frame are equivalent (Section 2.1). That is to say, there is no way to distinguish between them provided that measurements do not extend beyond the region of uniformity. One conclusion that will be drawn from this principle is that gravitational fields affect electromagnetic radiation: light leaving a star is red shifted, and light passing near a star is deviated from a classical straight-line path (Section 2.3). These effects and their experimental verification are also described.

2.1 The equivalence principle

The origin of the equivalence principle goes back to the experiments of Galileo. When Galileo compared the rate of fall of different materials he was attempting to answer a fundamental question: he wanted to know whether the gravitational attraction on different materials was the same. Even now it is not at all obvious that it should be so. Matter is constructed from very different particle species and the proportions of these vary from material to material. Each atom contains a nucleus which is made from nucleons (i.e. neutrons and protons) with electrons circulating around the nucleus. Nucleons feel the strong nuclear force whereas electrons do not; thus it is reasonable to ask whether nucleons and electrons feel the same gravitational attraction. The nucleon-to-electron ratio varies from unity for hydrogen to about 2.5 for elements with high atomic number, so that any difference in the gravitational force felt by nucleons and electrons would appear as a difference in the gravitational acceleration of elements of high and low atomic number.

Such a hypothetical difference would be tiny because the electron mass is only 1/2000th of the nucleon mass.

Another significant feature of matter is that the nuclear mass is smaller than the sum of the nucleon masses by the binding energy due to the nuclear force between them. This nuclear binding energy is zero for hydrogen where the nucleus is a single proton and rises to 0.7 per cent of the mass $\times c^2$ of the constituent nucleons in the case of iron. Hence if the gravitational force depended, like the strong force, on the number of nucleons rather than mass there would be a difference of 0.7 per cent between the gravitational acceleration of iron and hydrogen.

The analysis of experiments like Galileo's proceeds as follows. The force acting on a mass m_g in a gravitational field g is

$$F = m_g g.$$

Then according to Newton's second law of motion the acceleration a of this mass is given by

$$F = m_i a.$$

A distinction is made here between the gravitational mass m_g, on which the gravitational force acts, and the inertial mass m_i. Inertial mass appears in the expressions for kinetic energy ($m_i v^2/2$) and momentum ($m_i v$), so that its definition can be made independent of any weighing process. Eliminating F from the last two equations gives the acceleration

$$a = (m_g/m_i)g.$$

Tests from Galileo's time up to the present reveal no variation in the rate of fall from material to material. Therefore m_g/m_i has the same value for all materials, and by choosing units appropriately we make this ratio equal to unity. Einstein interpreted this result as follows: the motion of a neutral test body released at a given point in space–time is independent of its composition. This statement is known as the *weak equivalence principle* (WEP).

Einstein next considered the implications of the equivalence principle for motion in free fall, that is to say motion under gravitational forces alone. One modern example is a space capsule in orbit around the earth; another is a capsule falling radially toward the Earth. What is essential for free fall is that the capsule is not powered and that the atmospheric drag is negligibly small. Alternatively the capsule might be drifting in the weak gravitational field of intergalactic space. Einstein posed a searching question for such systems which we put in modern dress: can an astronaut inside the capsule determine his state of motion without looking out of the capsule? When the astronaut is in a uniform gravitational field he is incapable of determining his motion by any mechanical means. If, for instance, he drops a ball it accelerates at the same rate as the capsule and will remain at rest relative to the capsule, whatever their shared acceleration. However, if the capsule is in a region where the gravitational field is *not* uniform he can detect his motion. To give a concrete example, consider the capsule to be falling radially toward the Earth. Then dropping not one, but two balls will be an effective strategy because the gravitational forces acting on them converge toward the centre of the Earth as in Fig. 2.1. The astronaut could measure the resultant movement of the balls towards each other, given a large enough capsule and a long enough time interval. Such effects owe their origin to gradients in the field and are called *tidal* effects. The weak equivalence principle can now be restated as follows to exclude tidal effects: the results of *local* mechanical experiments in a state of free fall are independent of the motion. 'Local' is a technical term used here to express the restriction to a region sufficiently small that the gravitational field is effectively uniform. Einstein then generalized this form of the equivalence principle to cover both electromagnetic and mechanical experiments. It becomes the *strong equivalence principle* (SEP) which states the following:

1 the results of all local experiments in a frame in free fall are independent of the motion;
2 The results are the same for all such frames at all places and all times.

In a nutshell, physics is the same in all freely falling frames. Notice that one frame in free fall can have a very different velocity and acceleration from another such frame. For example we can compare the frames of satellites in free fall around different stars. The SEP can be viewed as a replacement or extension of the first postulate of SR. SR requires that the result of an experiment is the same for all inertial frames; the SEP requires that the result of local experiments be the same in all freely frames. In the special theory of relativity it is assumed that a single inertial frame can be applied to the whole Universe, but at the cost of neglecting acceleration! In the general theory of

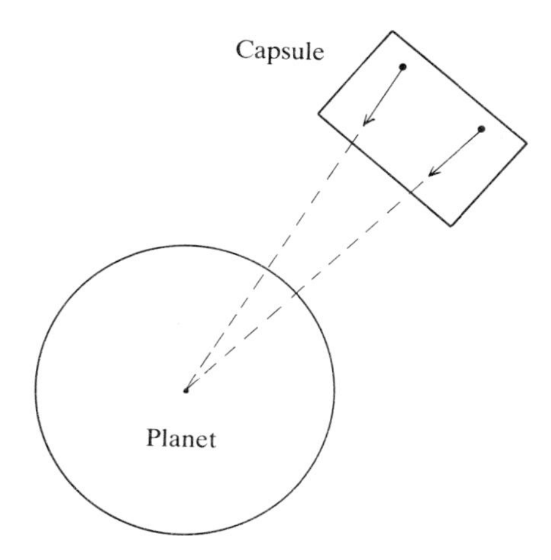

Fig. 2.1 The trajectories of objects in free fall within a space capsule.

relativity the natural frame anywhere is chosen to be a frame in free fall, but we cannot cover the whole Universe with one frame. Space–time as a whole is then seen to be a patchwork of such freely falling frames. The vast body of experimental data which has been accumulated in support of SR can be reconciled with the SEP by making the inference, as Einstein did, that physics in free fall must be consistent with SR. Thus we should add a rider to the SEP:

3 the results of local experiments in free fall are consistent with SR.

2.2 Experimental comparisons of m_g/m_i

The equivalence principle is so central to the development of GR that very refined tests have been carried out to check the equality of m_g/m_i for different materials. An experiment of Roll, Krotkov, and Dicke (1964), which compared the gravitational attraction of the Sun on different materials, is extremely precise. The apparatus consisted of a delicate torsion balance from which were hung masses made of the two materials under comparison (e.g. 30 mg masses of gold and aluminium). The balance had a triangular frame to reduce the effect of gradients in the gravitational field. Here for simplicity we shall take the frame to be a single arm suspended at its centre as shown in Fig. 2.2; Fig. 2.2(a) shows a view from above the North Pole, and Fig. 2.2(b) shows a side view. The forces acting on the 30 mg masses along the Sun–Earth axis are the gravitational attraction of the Sun ($\propto m_g$), the centrifugal force due to motion around the sun ($\propto m_i$), and a pull from the balance (F). We first write the equation for the balance of forces on the Earth itself:

$$\frac{GMm_g(\mathrm{e})}{R^2} = \frac{m_i(\mathrm{e})v^2}{R}$$

where M is the Sun's mass, $m(\mathrm{e})$ is the mass of the Earth, v is its orbital speed and R is the Earth–Sun separation. Simplifying this gives

$$v^2 = \frac{GM}{R}r(\mathrm{e})$$

where $r(\mathrm{e}) = m_g(\mathrm{e})/m_i(\mathrm{e})$. The equation for the balance of forces acting on the gold mass is similar:

$$F(\mathrm{Au}) + \frac{GMm_g(\mathrm{Au})}{R^2} = \frac{m_i(\mathrm{Au})v^2}{R}$$

where the additional term F is the pull from the arm of the balance. Then

$$F(\mathrm{Au}) = \frac{GMm_i(\mathrm{Au})}{R^2}[r(\mathrm{e}) - r(\mathrm{Au})],$$

with a corresponding expression for the aluminium mass. To simplify matters

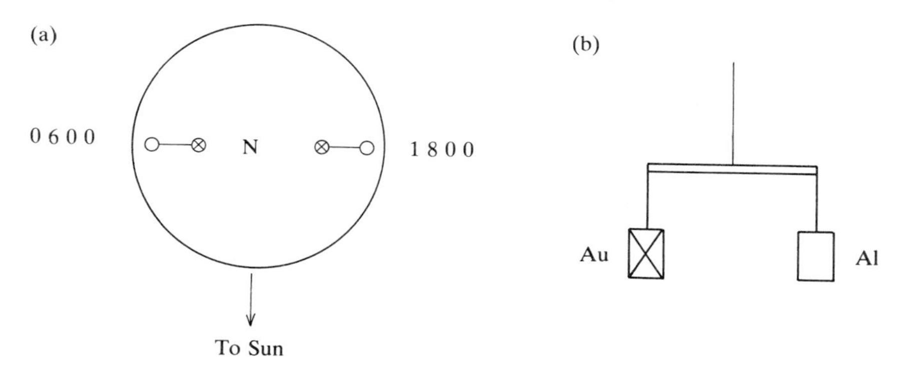

Fig. 2.2 The experiment of Roll *et al.* (1964): (a) view from above the North Pole; (b) side view.

we assume that

$$m_i(\mathrm{Au}) = m_i(\mathrm{Al}) = m$$

and that the arms of the balance are of equal length l. At 0600 hrs the torque on the balance is

$$\Gamma = [F(\mathrm{Au}) - \mathrm{F(Al)}]l = \frac{GMml}{R^2}\Delta$$

where $\Delta = r(\mathrm{Au}) - r(\mathrm{Al})$. Only if Δ is non-zero will there be a net torque, and this torque would be reversed 12 h later at 1800 h. Roll, Krotkov, and Dicke searched for any sign of a 24 h oscillation of the torsion balance which such a time-varying torque would stimulate, but with a null result. The apparatus was operated remotely in a vacuum and in a temperature-controlled chamber. All magnetic material was excluded, and the insulating elements of the balance were plated with metal to eliminate any electrostatic effects. With such care it was possible to obtain an upper limit of

$$\Delta \leqslant 3 \times 10^{-11}$$

The way that the rotation of the Earth is put to use in reversing the torque is a key feature. Earlier experiments had compared the Earth's gravitational attraction with the centrifugal force from its own rotation. The forces are larger than in the experiment of Roll, Krotkov, and Dicke but the torque can only be reversed by bodily rotating the apparatus, which gives large and irreversible effects that vitiate the method. They note that for a deviation Δ of 10^{-11} the accumulated effect over a year would only amount to a relative velocity of 1 $\mu m\ \mathrm{s}^{-1}$ towards the Sun!

The experiment of Roll, Krotkov, and Dicke does not entirely close the door on the possibility that gravitational and inertial mass are not equal. It has been proposed that the gravitational force has a short-range component that varies from material to material, with a suggested range of about 10^6 m or less. At the distance of the Earth from the Sun this force component would not contribute and would not be detectable in the experiment. One form proposed for the total potential, including the gravitational potential, is

$$\varphi = -\left(\frac{Gm_i m_j}{r}\right)\left[1 - B_i B_j \zeta \exp\left(\frac{-r}{\lambda}\right)\right]$$

for masses m_i and m_j a distance r apart. B_i and B_j are factors which make the mass per nucleon exactly one atomic mass unit; ζ specifies the strength of the new potential relative to gravitation and λ is its range parameter. The hypothetical new force has been called the fifth force. Its presumed dependence on the number of nucleons mimics the behaviour of the other three forces of nature (weak nuclear force, strong nuclear force, and electromagnetism) in showing a dependence on the number of particles rather than on mass. At the surface of the Earth, which is assumed to be a sphere of radius R and uniform density ρ, the relative acceleration of two objects towards the Earth as a result of such a fifth force would be

$$\Delta g \approx 3g\left(\frac{\zeta\lambda}{R}\right)\frac{\Delta(B)}{2}$$

if $\lambda \ll R$ and $|\zeta| \ll 1$; g is the usual gravitational acceleration and $\Delta(B)$ is the difference between the values of B for the two bodies. Experiments performed since 1986 appear to rule out any fifth force. For example, Kuroda and Mio (1989) have measured the relative acceleration in free fall of pairs of test bodies: aluminium/copper and aluminium/carbon. The two bodies labelled A and B in Fig. 2.3 carry corner cube mirrors C_A and C_B while B also carried a beam splitter S. These optical components are used as elements of a Michelson interferometer; the source is a stabilized He–Ne laser and the detector is a p-i-n photodiode. If the velocity of fall of object A is V_A, then the frequency ν of the light reflected from A undergoes a Doppler shift to a frequency

$$\frac{\nu(c - V_A)}{c + V_A} \approx \frac{\nu(1 - 2V_A)}{c},$$

with a similar expression for light reflected from B. These reflected beams interfere at the detector giving beats of frequency $(V_A - V_B)\nu/c$ in the amplitude. Therefore the intensity of the photodiode signal is modulated at twice this frequency:

$$\Delta\nu = \frac{2(V_A - V_B)\nu}{c}.$$

If there is no relative acceleration between A and B, then $\Delta\nu$ remains constant during the time that A and B are in free fall. This fall is over 60 cm and takes

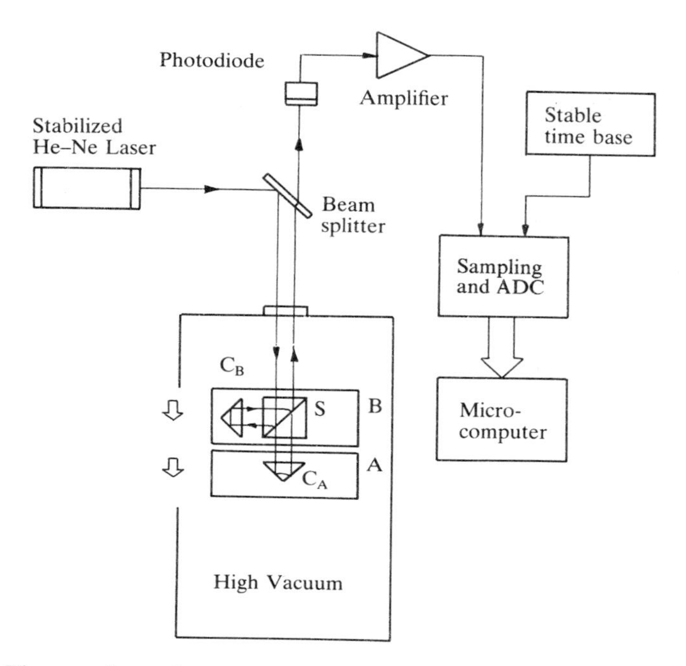

Fig. 2.3 The experimental apparatus of Kuroda and Mio (1989). (Courtesy Physical Review Letters and the authors.)

0.35 s. The output from the photodiode is amplified and sampled by an analogue-to-digital convertor (ADC) every 4 μs, and these values are stored in a microcomputer memory. A and B are made in the form of a saddle and rider so that their centres of mass lie within 0.3 mm on release. Each weighs about 0.1 kg. Magnetic material is excluded and the bodies are earthed before release. Residual gas (2×10^{-10} atms) produces a repulsive force between A and B. There is also a mutual gravitational attraction between A and B. These effects are calculated to produce relative accelerations of order 3×10^{-9} m s^{-2}. After correction for all such effects the observed acceleration difference found for aluminium/carbon is $-0.18 \pm 1.38 \times 10^{-8}$ ms^{-2}; which is a null result at the level of precision achieved in the experiment. Experimental tests are therefore fully consistent with the equality of inertial and gravitational mass.

2.3. The gravitational spectral shift and the deflection of electromagnetic radiation

The first important consequence of the SEP pointed out by Einstein is the prediction that gravitational fields affect radiation. Consider a capsule of height Δr in free fall towards the Earth with acceleration g. A pulsed light source is located on the floor of the capsule. According to the SEP light will travel at a velocity c in any direction in the frame of the capsule, so that light will reach the roof after a time $\Delta r/c$. If the capsule is assumed to start from rest at the moment that the light flash is emitted, then the capsule will have attained a velocity

$$v = g\Delta r/c$$

relative to the Earth when the light reaches the roof. An external observer at rest who detects the flash at the moment that it reaches the roof would see it Doppler shifted to a lower frequency than observed in the source frame. The effect of the relative velocity v of the source frame away from the external observer is to give a fractional change in frequency (to first order in v/c)

$$\Delta \nu/\nu = v/c = g\Delta r/c^2.$$

Note that the relevant velocity is that of the source frame relative to the external observer at the instant that the light is detected.

The crests of the waves emitted by the source can be regarded as the ticks of an atomic clock, and if the rate of ticking ν increases, the same time interval t would contain more ticks. Hence the frequency shift implies that time intervals measured in the two frames by identical clocks would differ, i.e. time is distorted. The interval t between ticks emitted by the source when it is at rest at height r is given by

$$\frac{1}{t(r)} = \nu.$$

The interval between ticks measured by the external observer also at rest, but at a height $r + \Delta r$, is given by

$$\begin{aligned} \frac{1}{t(r + \Delta r)} &= \nu - \Delta \nu \\ &= \nu\left(1 - \frac{g\Delta r}{c^2}\right) \\ &= \nu\left(1 - \frac{GM\Delta r}{r^2c^2}\right) \end{aligned}$$

where M is the Earth's mass. Thus to a first-order approximation

$$t(r + \Delta r) = t(r)\left(1 + \frac{GM\Delta r}{r^2c^2}\right).$$

By integrating this last equation from r to infinity the time interval measured at a remote point where the gravitational potential is negligible can be compared with the same time interval measured at r:

$$t(r) = t(\infty)\left(1 - \frac{GM}{rc^2}\right).$$

Re-expressing this result in terms of the gravitational potential $\varphi = -GM/r$ gives

$$t(\varphi) = t(0)\left(1 + \frac{\varphi}{c^2}\right).$$

Notice that because the gravitational potential is always negative $t(0)$ is always larger then $t(\varphi)$. The time interval $t(0)$ measured remotely is called the co-ordinate time and the time $t(\varphi)$ measured where the gravitational potential is φ is called the local proper time. The remote observer measures time intervals to be dilated and light to be red shifted. Later we shall need to relate the squares of time intervals in different frames, and so we write

$$\mathrm{d}t^2(r) = \mathrm{d}t^2(\infty)\left(1 - \frac{2GM}{rc^2}\right). \tag{2.1}$$

The derivation leading to eqn (2.1) is only valid for low velocities and only accurate to order $2GM/rc^2$. However, eqn (2.1) itself is taken to be an exact result in GR. We rewrite eqn (2.1) for later use as

$$\mathrm{d}\tau^2 = \mathrm{d}t^2\left(1 - \frac{2GM}{rc^2}\right) \tag{2.2}$$

where $\mathrm{d}\tau$ is the proper time and $\mathrm{d}t$ is the coordinate time.

A precise measurement of the gravitational spectral shift was made by Pound and Rebka (1960) using photons that travelled the length of a tower 22.6 m tall at Harvard University. The predicted spectral shift is only

$$\Delta\nu/\nu = 2.46 \times 10^{-15}.$$

Pound and Rebka exploited the contemporary discovery by Mössbauer that the linewidths of certain gamma emitters, in particular $^{57}\mathrm{Fe}^*$, were exceptionally narrow and much closer to the resolution required to detect the gravitational red shift than for any other type of source. The distinguishing feature of a Mössbauer transition is that, whereas in a normal gamma decay the nucleus recoils against the photon, in a Mössbauer transition it is the crystal lattice as a whole that recoils against the photon. In a normal decay the photon and the nucleus share the energy release Q, so that the photon energy is less than Q. Similarly, when a photon is absorbed in the reverse process

$$\gamma + {}^{57}\mathrm{Fe} \rightarrow {}^{57}\mathrm{Fe}^*,$$

the nucleus recoils to take up the photon momentum and so acquires some kinetic energy. Thus the photon must have energy greater than Q if it is to be absorbed. Suppose that in a decay the recoil momentum is p; then the kinetic energy of the nucleus is $p^2/2M$ where M is its mass. It follows that in a Mössbauer transition the recoil energy is drastically reduced because M is now the mass of the crystal. Consequently the laboratory energy of photons in emission and for absorption become effectively identical. A second important consequence of the Mössbauer effect is that the impact of thermal excitation on the width of the emission and absorption lines is reduced. Any thermal energy is shared by the whole lattice rather than a single nucleus. As a result, in the Mössbauer transition a $^{57}Fe^*$ source emits 14.4 keV photons with a fractional linewidth of only 10^{-12}. However, this is still about 500 times larger than the gravitational red shift that is to be measured.

Pound and Rebka placed the $^{57}Fe^*$ source at one end of the tower and a thin ^{57}Fe absorber at the other end. The absorber covered a scintillator which was viewed by a photomultiplier. It was arranged that the source could be driven slowly to and fro at a low velocity using a transducer; this motion produced a small Doppler shift in the frequency and energy of the emitted photons. The Doppler shift was used to compensate for the gravitational spectral shift of photons travelling along the tower. When the compensation was exact the absorption of the photons by the thin layer of ^{57}Fe was maximized and the counting rate on the photomultiplier behind it was minimized. Pound and Rebka were able to scan across the very narrow line profile by varying the source velocity and in this way achieved a large gain in sensitivity for locating the line centre. Their procedure enabled them to make a very precise measurement of the gravitational spectral shift:

$$\Delta\nu/\nu = (2.57 \pm 0.26) \times 10^{-15}$$

which agrees, to within the small quoted error, with the prediction obtained from the SEP. Other more recent tests involve direct comparison of the time-keeping of atomic clocks or of masers. Vessot *et al.* (1980) compared the rate of a hydrogen maser launched to a height of 10 000 km in a rocket with the rate of an identical maser kept in the laboratory. Two-way telemetry was used to compensate both for the atmospheric effects and for the first-order Doppler shift due to the relative velocity of the masers. The comparison gave a rate difference that agreed with the prediction from the SEP to parts in 10^4.

An extension of the above arguments based on the SEP leads to the conclusion that light can be bent in a gravitational field. We start with the observation that a photon climbing a distance d against the Earth's gravitational pull loses energy

$$\Delta E = h\Delta\nu = -gd\frac{h\nu}{c^2}$$

where h is Planck's constant. The kinetic energy lost by a body of mass m rising through the same distance is remarkably similar:

$$\Delta E = -gdm.$$

Therefore it appears that a photon of energy $E = h\nu$ behaves in a gravitational field as if it possessed an inertial mass E/c^2! To this extent, then, a photon feels the gravitational force and it can be argued that a photon should travel along a curved path in a gravitational field. This inference is reinforced by the following thought experiment. Imagine a space capsule in free fall near the Earth: inside it an astronaut strapped to one wall shines a beam of light horizontally at the opposite wall. Figure 2.4(a) shows the light path as seen by the astronaut; his frame is in free fall so that the light travels in a straight line to the opposite wall. An external observer at rest sees things quite differently. At emission the lamp is at one height, but by the time the light reaches the other wall the capsule has fallen a little. This view is shown in Figure 2.4(b) where the light path is seen to curve. Einstein calculated the deviation of starlight passing near the Sun's surface on its way to the Earth;

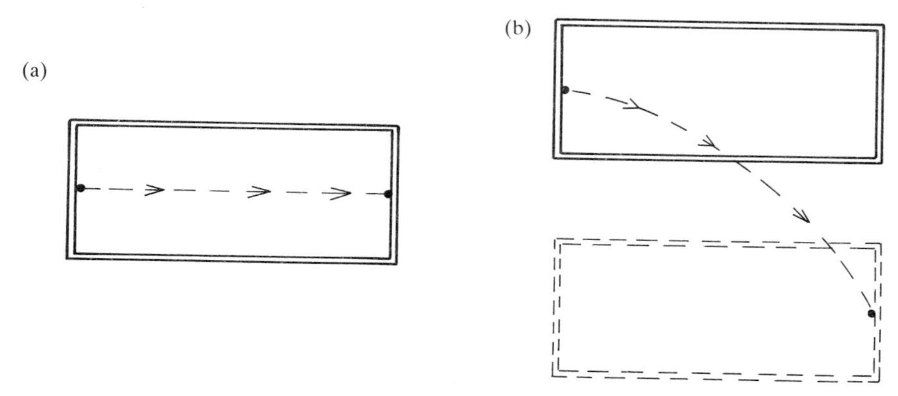

Fig. 2.4 Light path in a space capsule in free fall near the Earth as seen by (a) an occupant of the capsule and (b) an external observer at rest with respect to the Earth's surface.

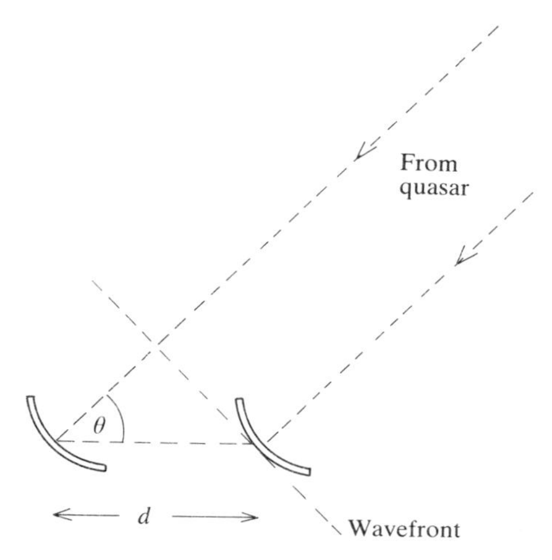

Fig. 2.5 The arrangement of a pair of radio antennas used to determine the angular position of radio-sources.

his result was 1.750 arcsec. We should note that a calculation using just the SEP gives exactly half this value; we shall look at the complete calculation in Chapter 8. The success of the 1919 measurement has already been discussed. Higher precision is obtained by studying the apparent motion of radio-sources that pass near to and behind the Sun's disc, a technique which is not restricted to times of solar eclipse. In Fig. 2.5 a widely spaced pair of radio-telescopes are shown receiving signals from the same source. If the source direction makes an angle θ with the baseline, which is of length d, then the path difference to the two dishes is $d \cos \theta$ and the phase lag between their signals is

$$\Delta(\text{phase}) = \frac{2\pi}{\lambda}(\text{path difference})$$

at a wavelength λ, and in terms of frequency ν this becomes

$$\Delta(\text{phase}) = \frac{2\pi \nu d \cos \theta}{c}.$$

Measurement of the phase difference leads to a determination of θ. Fomalont and Sramek (1976) used radio-telescopes on a 35 km baseline at Green Bank, West Virginia, to study radio-sources close to the Sun at frequencies of 2.7 and 8.1 GHz. This is an example of very long baseline interferometry (VLBI). They found a mean deviation due to the Sun's gravitational field corresponding to 1.760 ± 0.016 arcsec at the Sun's limb, which is in excellent agreement with Einstein's prediction.

3
Space curvature

The experimental observation of the gravitational spectral shift and of the bending of light passing close to the Sun indicate that space–time is distorted by the presence of matter. In Section 3.1 the basic concepts of curvature are introduced, based on two-dimensional examples. The fundamental point to be grasped is that the inhabitants of a curved space can determine its curvature from intrinsic measurements, i.e. from measurements confined to the space itself. All the geometrical information about a space is quantified by the equivalent of Pythagoras' theorem which relates the distance between nearby points in space to their coordinate separations. This fundamental equation is known as the metric equation. Straight lines, in the usual sense, cannot be drawn on curved surfaces or in curved spaces. Their place is taken by geodesics which are the straightest lines that can be drawn between a pair of points in a curved space and as such have no component of curvature in that space. The definition of curvature and ways to measure it are described in Section 3.2. Local vectors and how to compare local vectors at different places in a curved space are the topics of Section 3.3. The relationships between curvature and the metric equation is described in Section 3.4. In Section 3.5 there is a discussion of three-dimensional spherically symmetric spaces, which prepares the way for the later treatment of the space–time curvature in isotropic homogeneous cosmological models.

3.1. Two-dimensional surfaces

The initial difficulty with discussing curvature in our three-dimensional world is best appreciated by referring to the analogous case of a two-dimensional being living in a curved two-dimensional surface. This being believes that he is able to look and measure in all directions, whereas his observations are really limited to directions contained in the two-dimensional surface. He would receive two-dimensional light signals that had travelled within the surface and he would not be directly aware of any curvature. From our three-dimensional vantage the curvature that eludes him is obvious. However, there are intrinsic properties of any curved space which enable beings confined to that space (as we are confined to three-space dimensions) to detect and measure that curvature. These properties were first appreciated and investigated by Gauss.

Two-dimensional spaces are particularly easy to visualize and so make an ideal vehicle for introducing the concepts that will be used later in discussing curvature in space–time (three spatial dimensions plus one time dimension). Figure 3.1 shows one two-dimensional surface, a spherical surface; and Fig. 3.2 shows another, a cylindrical surface. Although both look curved there exists a fundamental difference between them. The cylindrical surface can be slit along its length (at AA) and unrolled to lie flat on a plane; however, there is no way that the sphere can be unrolled so that it lies flat. Maps of the whole Earth show that this can only be achieved by stretching some parts of the surface and shrinking others. A related property of the cylinder is that if the shortest paths between pairs of points, such as BC, are drawn on the surface then these become straight lines when the cylinder is unrolled. The cylinder is therefore called intrinsically flat, although not planar, and the spherical surface is intrinsically curved. Here we ignore topological subtleties in the case of the cylinder by considering only paths that do not cross the cut.

Suppose that rectangular Cartesian coordinates are drawn to cover a rectangular sheet of paper and it is then rolled up to form a cylinder. Distances s measured over the surface between a pair of points whose coordinate separations are x and y will be given by Pythagoras' theorem

$$s^2 = x^2 + y^2.$$

By contrast, an attempt to construct a set of rectangular Cartesian coordinates to cover the spherical surface runs into irretrievable difficulties. Taking the example of the Earth's surface, which we assume to be a perfect sphere, we could place the origin at the North Pole N as in Fig. 3.1. The x axis runs toward Quito (Q, along longitude 80° W) and the y axis runs toward Libreville (L, along longitude 10° E); both cities lie close to the Equator. On travelling along the x axis to Quito a traveller would find that Libreville lies in the y direction. If instead the traveller goes along the y axis to Libreville, Quito is found to lie in the x direction. Such inconsistencies make it impossible to cover a spherical surface using Cartesian coordinates. Although this is so, it does not trouble anyone surveying a piece of land. Cartesian coordinates work well enough for regions with linear dimensions small relative to the radius of the surface. We say that locally the surface is Euclidean and distances are given by the *differential* Pythagoras equation

$$\mathrm{d}s^2 = \mathrm{d}x^2 + \mathrm{d}y^2,$$

where $\mathrm{d}x$, and $\mathrm{d}y$ are the coordinate separations of nearby points on the surface.

One set of coordinates that can be used to cover the surface of the whole sphere are the polar angles (θ, φ). With the origin at the centre of the Earth θ runs from 0° at the North Pole to 180° at the South Pole, and is related to the latitude. φ is essentially the longitude and runs from $-180°$ to $+180°$. Locally, that is to say on a scale small compared with the radius of curvature r of the surface, the distance between two points at (θ, φ) and $(\theta + \mathrm{d}\theta, \varphi + \mathrm{d}\varphi)$

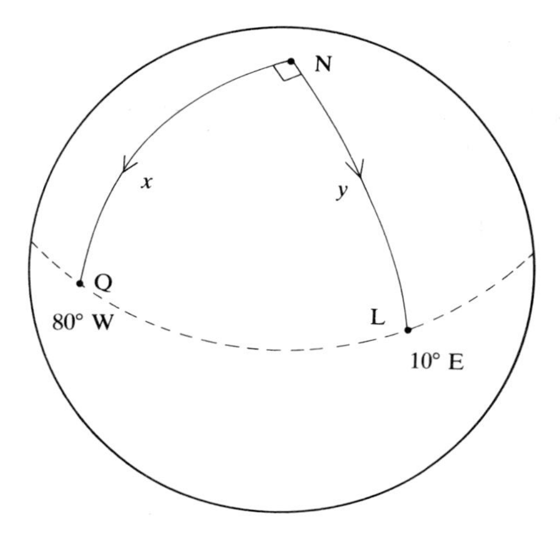

Fig. 3.1 A two-dimensional spherical surface, such as the surface of the Earth.

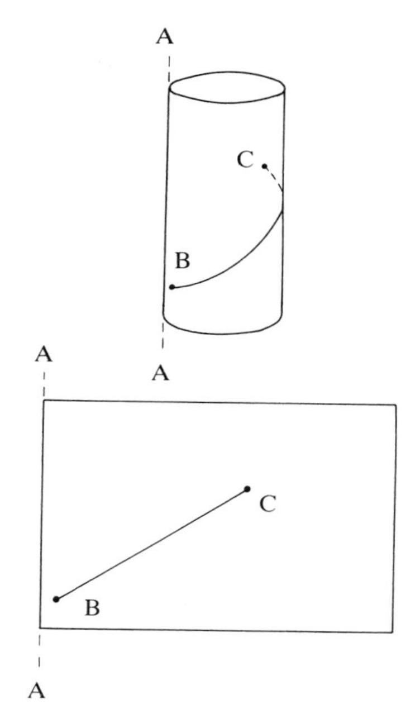

Fig. 3.2 A two-dimensional cylindrical surface. In the lower diagram the cylinder has been unwrapped onto a flat surface.

is $r\,d\theta$ in latitude and $r \sin\theta\,d\varphi$ in longitude. Then the total separation is given by the quadratic equation

$$ds^2 = r^2\,d\theta^2 + r^2 \sin^2\theta\,d\varphi^2$$

This equation and the preceding one are called *metric equations.* It is easy to see that the metric equation is an essential property of a surface. For, suppose an attempt is made to reconstruct the geometry of the Earth using only the information available in the index of an atlas. Each town appearing in the index has two coordinates (latitude and longitude) which range from -90 to $+90$ and from -180 to $+180$ respectively. If these locations are plotted on a sheet of paper treating the coordinates as Cartesian coordinates the result would not resemble the Earth's surface. A distance scale is needed in interpreting the coordinates and this is what the metric equation provides. For example, the metric equation above indicates that near the poles ($\theta = 0°$ or $180°$) changes in longitude φ produce only small displacements.

The properties of a spherical curved surface which have been highlighted above are common to other curved two- and higher-dimensional spaces of interest. First of all, in order to cover the whole of a curved surface or space it is necessary to use generalized or Gaussian coordinates; rectangular Cartesian coordinates are inadequate. A second property is that the local distances are given by a metric equation in the Gaussian coordinate separations. The spaces of interest to us here belong to a category called *Riemann spaces* which are distinguished by having metric equations that are quadratic in the coordinate separations.

We now introduce a key property of Riemann spaces, namely that it is always possible to match any arbitrary region of such a space by a flat space provided that the region is taken small enough. This is rather like being able to draw a straight line tangent at any point on a smooth curve. A first step is to consider a two-dimensional Riemann surface with metric equation

$$ds^2 = g_{11}\,dv^2 + 2g_{12}\,dv\,dw + g_{22}\,dw^2$$

where (v, w) are some Gaussian coordinates. The coefficients g_{11}, g_{12} and g_{22} are functions of position and contain all the information about the geometry of the surface. g_{11} is taken to be positive (which can be achieved by multiplying throughout by -1 if necessary). A point P on the surface is selected and coordinates (x, y) are found for which the metric equation is locally Euclidean around P. A general definition of new coordinates is

$$dv = A(x, y)\,dx + B(x, y)\,dy$$

$$dw = C(x, y)\,dx + D(x, y)\,dy,$$

where

$$A = \partial v/\partial x, \qquad B = \partial v/\partial y, \qquad C = \partial w/\partial x \qquad \text{and} \qquad D = \partial w/\partial y.$$

Then

$$ds^2 = g'_{11}\,dx^2 + 2g'_{12}\,dx\,dy + g'_{22}\,dy^2;$$

where

$$g'_{11} = A^2 g_{11} + 2AC g_{12} + C^2 g_{22},$$

$$g'_{12} = AB g_{11} + AD g_{12} + BC g_{12} + CD g_{22}$$

and

$$g'_{22} = B^2 g_{11} + 2BD g_{12} + D^2 g_{22}.$$

We are free to choose not only the values of A, B, C, and D at P, but also the values of their derivatives at P. The values of A, B, C, D, and their six independent first derivatives provide ten free variables. There is just enough flexibility to arrange that at P

$$g'_{11} = g'_{22} = +1, \qquad g'_{12} = 0;$$

and in addition that the first derivatives of the metric components vanish

$$\partial g'_{11}/\partial x = \partial g'_{12}/\partial x = \partial g'_{22}/\partial x = 0,$$

$$\partial g'_{11}/\partial y = \partial g'_{12}/\partial y = \partial g'_{22}/\partial y = 0.$$

It follows that a Euclidean surface with metric equation

$$ds^2 = dx^2 + dy^2$$

will match the curved surface locally at P. In other words a plane can always be drawn at any arbitrary point on a two dimensional Riemann surface so that it is locally tangential to the surface. This result agrees with our experience for surfaces embedded in three space. Notice that the conditions on the metric components and derivatives only make up nine equations, whereas there are ten degrees of freedom. The residual degree of freedom amounts to the choice of orientation of the x and y axes on the plane. A similar procedure can be followed in higher dimensional spaces. Some coordinate transformation can always be found which converts the metric equation to a sum of squares. The Cartesian coordinates which result from this transformation describe a space which is tangential to the curved space at the point selected. Because a flat tangent space can always be drawn locally to any point in a Riemann space, Riemann spaces are said to be locally flat (or locally Euclidean). It is not possible to arrange that the second derivatives as well as the first derivatives of the coefficients g_{11}, g_{12} and g_{22} all simultaneously vanish, because there are insufficient free variables.

One point which we have chosen to overlook so far is highly relevant. The original metric equation can be rewritten

$$ds^2 = \left(g_{11}^{1/2}\,dv + \frac{g_{12}\,dw}{g_{11}^{1/2}}\right)^2 + \left(g_{22} - \frac{g_{12}^2}{g_{11}}\right) dw^2$$

from which we can obtain

$$ds^2 = dx^2 + dy^2$$

by making the substitutions

$$dx = g_{11}^{1/2}\, dv + \frac{g_{12}\, dw}{g_{11}^{1/2}} \qquad \text{and} \qquad dy = \left(g_{22} - \frac{g_{12}^2}{g_{11}}\right)^{1/2} dw$$

This presumes that $(g_{11}g_{22} - g^2{}_{12})$ in the original metric equation is positive. On the other hand if $(g_{11}g_{22} - g^2{}_{12})$ is negative the quadratic form obtained is

$$ds^2 = dx^2 - dy^2.$$

The space involved is still locally flat as it has a metric equation that reduces to a difference of squares. Its tangent space is called pseudo-Euclidean and the space iself is called pseudo-Riemannian. For most purposes we can ignore the distinction and include the pseudo-Riemann spaces with the Riemann spaces. Referring back to Section 1.1, we can see that the space–time of SR is pseudo-Euclidean, which indicates the reason for our interest in such spaces.

The shortest paths joining points a finite distance apart on a curved surface are generally not simple straight lines. However, they are the straightest lines that can be drawn on the surfaces between the points under discussion. Such paths are called *geodesics* and naturally on a flat surface a geodesic is a straight line. Geodesics on a spherical surface are the well known great circle routes which are frequently used by airlines on intercontinental flights. Figure 3.3(a) shows a great circle path drawn to touch a line of latitude θ at P;

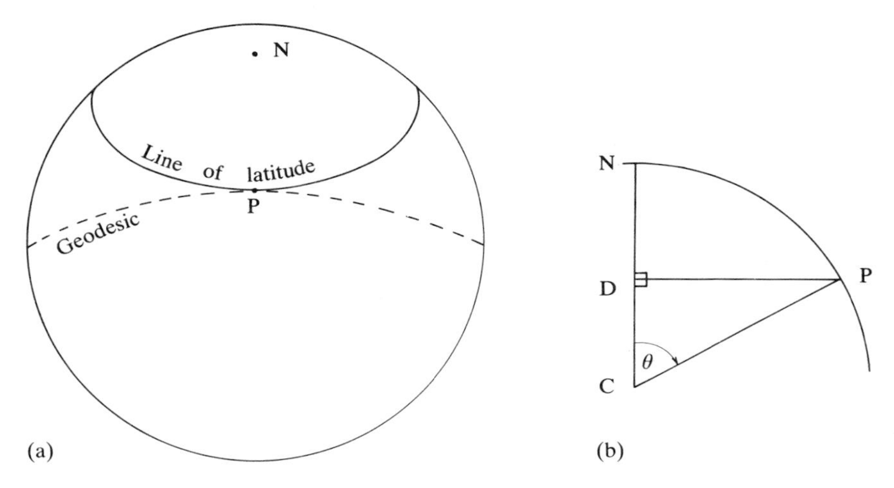

Fig. 3.3 (a) A line of latitude and a geodesic (great circle) on a spherical surface which are tangential to one another at P; (b) a diametral plane section through the sphere which contains the North Pole N and the centre of the Earth C. (D is the centre of curvature of the line of latitude at P).

Fig. 3.3(b) shows a diametral plane section of the sphere through P and the North Pole N. D is the centre of curvature of the line of latitude while C, the centre of the sphere, is the centre of curvature of the geodesic. The geodesic has curvature $1/r$ and the line of latitude has curvature $1/(r \sin \theta)$. Resolving the curvature of the line of latitude perpendicular and parallel to the surface gives $(1/r \sin \theta) \sin \theta = 1/r$ and $\cos \theta/r \sin \theta$; while for the geodesic great circle the corresponding components are $1/r$ and 0. This illustrates a very important property of any set of lines through a point on a surface which are tangential (share the same direction) at that point. The component of curvature perpendicular to the surface is the same for all of them, and in the case illustrated this is $1/r$. Among these lines one, the geodesic, has only this component of curvature. It has no component of curvature lying in the surface, which makes it the straightest path possible over the surface. This characteristic is quite general and holds for geodesics in spaces that are less symmetric than that of a sphere, and in spaces of higher dimension.

3.2. Measurements of curvature

Intrinsic methods by which an observer in two dimensions can make qualitative and quantitative measurements on the curvature of the surface he inhabits will be described next. A quantity called the Gaussian curvature quantifies the local curvature of any two-dimensional surface and can be generalized in the case of higher dimensions. One simple measurement reveals whether a two-dimensional surface is curved and determines the sign of curvature. The observer pegs one end of a string of length r to a fixed point O and then makes a circuit round O so that the string is all the while fully stretched, but of course confined to lie in the surface. He then measures the circumference C of one complete circuit around O. The result for the flat surface F drawn in Fig. 3.4 is

$$C_F = 2\pi r.$$

However, if the surface is dome-shaped like the surface P the circumference is shorter and

$$C_P < 2\pi r.$$

In the case of the saddle-shaped surface N shown in Fig. 3.4, the circumference is longer and

$$C_N > 2\pi r.$$

Now P is a surface of positive curvature like a spherical surface, while N has negative curvature, and so the difference $2\pi r - C$ fixes the *sign* of curvature. The reader will find it hard to picture the surface N continuing everywhere with equal negative curvature. The reason is that surfaces of negative curvature cannot be imbedded into our (nearly) flat and Euclidean three-space. The inhabitant of two dimensional space could also refine the use of his measure-

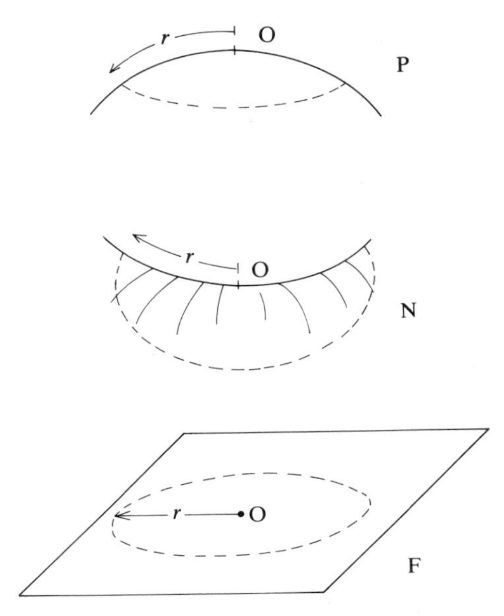

Fig. 3.4 Three two-dimensional surfaces: P is dome shaped, N is saddle shaped, and F is flat. In each case the broken line marks out a path which stays exactly a distance r from O.

ment to make a quantitive determination of curvature. We specialize to the case of a spherical surface whose cross-section is shown in Fig. 3.5. C is the centre of curvature and R is the radius of curvature. The angle subtended by a string of length r at the centre of the sphere is θ, and so

$$\theta = r/R$$

where R is the radius of the sphere. Then

$$C_P = 2\pi R \sin\theta$$

i.e.

$$C_p = 2\pi r\left(1 - \frac{r^2}{6R^2} + \cdots\right).$$

In the limit as r tends to zero we obtain

$$R^{-2} = \frac{3}{\pi}\lim_{r\to 0}\left(\frac{2\pi r - C_P}{r^3}\right).$$

This procedure certainly provides an intrinsic method for measuring the radius of a spherical surface. In fact it provides more because it yields the curvature K of any two-dimensional surface. For the case of the sphere K is simply $1/R^2$. In order to extend the discussion beyond the sphere we must introduce a definition of curvature valid for any such surface.

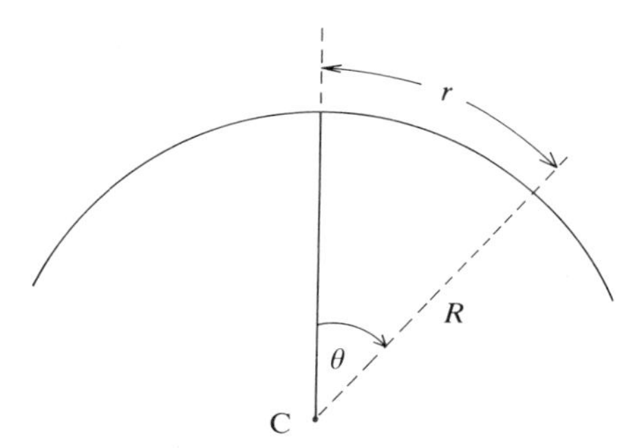

Fig. 3.5 A section through a spherical surface, with centre of curvature at C and radius R.

A portion of a more general two-dimensional surface is pictured in Fig. 3.6. G_1OG_1 and G_2OG_2 are geodesics across the surface which intersect at right angles; ON is the local normal to the surface. Rectangular Cartesian coordinates can be assigned at O using the tangents to G_1OG_1 and G_2OG_2 plus ON itself. These will be called the v, w, and z axes respectively. An expression for z which is valid for a point (w, v, z) on the surface close to O such that both w and v are small can be obtained by a Taylor series expansion:

$$z = \frac{\partial z}{\partial v}v + \frac{\partial z}{\partial w}w + \frac{1}{2}\left(\frac{\partial^2 z}{\partial v^2}v^2 + 2\frac{\partial^2 z}{\partial v \partial w}vw + \frac{\partial^2 z}{\partial w^2}w^2\right) + \cdots$$

The first two terms vanish because the v, w plane is parallel to the surface at O. Therefore to second order in v and w we have

$$z = \frac{1}{2}(Lv^2 + 2Mvw + Nw^2).$$

This expression can be converted to a sum of squares simply by rotating the Cartesian frame around ON through an angle

$$\frac{1}{2}\tan^{-1}\left(\frac{2M}{L - N}\right).$$

In the new Cartesian frame with coordinates (x, y, z) the surface has the equation

$$z = \tfrac{1}{2}(K_1x^2 + K_2y^2).$$

Now consider the line along which this surface intersects the plane xOz. It has the equation

$$z = K_1x^2/2,$$

and its radius of curvature R_1 is given by the formula relating the sagitta z to the chord length $2x$ for a circular arc:

$$2R_1z = x^2.$$

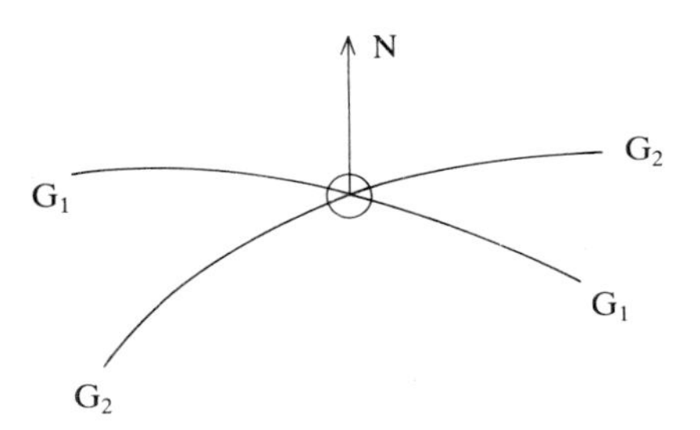

Fig. 3.6 G_1OG_1 and G_2OG_2 are geodesics on a two-dimensional surface which intersect at right angles. ON is the normal to the surface at O.

Thus the curvature of the line of intersection is

$$1/R_1 = K_1.$$

Similarly in the orthogonal section with yOz the curvature is K_2. It is not too difficult to prove that K_1 and K_2 are the minimum and maximum curvatures for any plane section through the surface at O containing ON. K_1 and K_2 are called the *principal curvatures* of the surface at O. Their product is an invariant for the surface at O called the *Gaussian curvature* K; thus

$$K = K_1 K_2. \tag{3.1}$$

A sphere has Gaussian curvature R^{-2} at any point on its surface. In the case of a cylinder one principal plane bisects its length along a straight line; hence one principal curvature is zero, and the Gaussian curvature is also zero. Finally, for the saddle-shaped surface of Fig. 3.4 one of the principal planes lies along the length of the saddle in the direction of the horse's spine, and the other lies transverse to the saddle in the direction of the horse's ribs. The centre of curvature of the first section lies above the saddle, whilst the centre of curvature of the second section lies below the saddle. Therefore K_1 and K_2 have opposite signs and the Gaussian curvature K is negative. The method already described for measuring the curvature of the sphere generalizes so that for any other two-dimensional surface the curvature is given by

$$K = \frac{3}{\pi} \lim_{r \to 0} \left(\frac{2\pi r - C}{r^3} \right).$$

Two other intrinsic methods of measuring K are worth discussing as they are methods that carry through to the analysis of space–time curvature in Chapter 7. The first method is based on the way that the separation of geodesics grows with distance. In the case of a plane the separation between a pair of geodesics (straight lines) through a point increases linearly with the distance from this point. In contrast the separation of the pair of lines of longitude shown diverging from the pole N in Fig. 3.7 does not vary linearly with the distance s measured from the pole. The difference in longitude is φ and so the separation after a distance s is

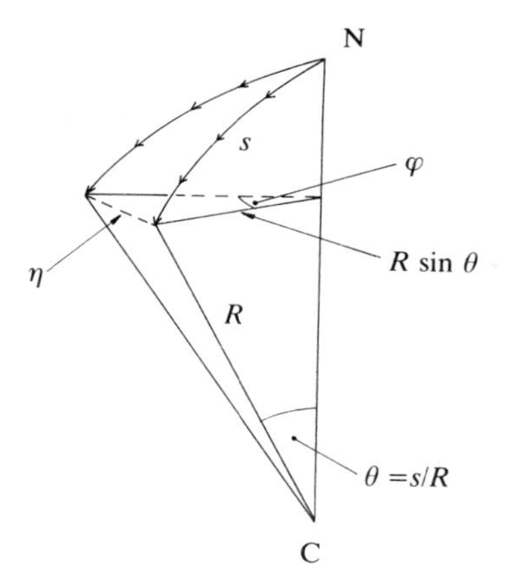

Fig. 3.7 A diagram to show how the separation of a pair of geodesics on a sphere depends on the distance from their intersection point.

$$\eta = (R \sin \theta)\varphi$$

where R is the radius of the sphere and

$$\theta = s/R.$$

Thus

$$\eta = (R\varphi) \sin(s/R).$$

Differentiating this expression gives

$$\frac{d^2\eta}{ds^2} = \frac{-\eta}{R^2}.$$

This relation generalizes for all two-dimensional surfaces to become

$$\frac{d^2\eta}{ds^2} = -\eta K \tag{3.2}$$

where K is again the Gaussian curvature. Another effect of intrinsic curvature is revealed when a vector is transported round a closed path in such a way that it is moved parallel to itself at each step of the journey.

3.3 Local vectors and parallel transport

Vectors are often drawn as extended lines with arrowheads. This convention requires careful interpretation for local vectors, i.e. vectors measured at a given point in space. Two examples will be considered. First, wind velocity is measured at a given observation point and refers solely to that point, for all that it may be convenient to show it on a chart as an arrow apparently extending for a long distance. In a like manner the force on an electron in an electric field has magnitude and direction but only exists at the coordinates of the electron. Wind velocity and force are both local vectors. A simple and most useful local vector in the present context is the tangent vector to a curve in space.

The comparison of local vectors at different places is easy in flat space. First a Cartesian coordinate system is set up with one vector $\boldsymbol{a}$ at a position A and the other $\boldsymbol{b}$ at position B; $\boldsymbol{a}$ and $\boldsymbol{b}$ will be equal if their components are equal. Equivalently we can imagine picking up $\boldsymbol{b}$ and carrying it to A without changing its length or direction and then examining whether it fits $\boldsymbol{a}$ exactly. In this procedure $\boldsymbol{b}$ is said to be parallel transported from B to A. Comparison of local vectors in a curved space is less straightforward because it is no longer possible to set up a single Cartesian coordinate system to cover all space. Once more it is helpful to consider the situation of a two-dimensional being who lives on a spherical surface embedded in our three-dimensional space. Suppose that he starts at the pole in Fig. 3.8 with the local vector $\boldsymbol{a}$ shown there. His only strategy is to take small steps and to carry the local vector parallel to itself over each step. Without any reference frame to check the parallelism even this seems difficult. However, he can start by moving off in the direction of the local vector itself, and in this case parallel transport is well defined. What the being is doing in this case is to trace out a geodesic—a great circle on the sphere, e.g. NA shown in Fig. 3.8. After this he can carry any other vector parallel to itself by travelling along a geodesic and keeping the local vector at a constant angle to the geodesic. In cases where the route does not follow a geodesic it would be necessary to split the path into infinitesimal steps. Then, because the space is locally flat, each such path element is part of some geodesic.

Emboldened by his success, the being could go on to parallel transport the vector $\boldsymbol{a}$ along the closed path NABN in Fig. 3.8. Each path segment is a geodesic. Starting from the pole the local vector is carried along the direction it points (a line of longitude) to the equator (A). From A it is parallel transported along the equator to B and then returned along another line of longitude to the pole. As a result the vector is seen to rotate through an angle φ which is the separation in longitude between A and B. In general the rotation of a local vector when carried round a closed path on any two-dimensional surface is given by the expression

$$\varphi = K \text{ (area enclosed by path)} \tag{3.3}$$

which is easily checked for the route discussed.

An equivalent view of this effect is that the result of parallel transport in a curved space depends on the path taken. Instead of carrying the vector $\boldsymbol{a}$ round the closed path NABN, the being could compare the result of carrying $\boldsymbol{a}$ to B along one path NAB with the result for the direct path NB. Again the difference in angle would be φ. The effect of parallel transporting a local vector round a closed path on a cylinder is shown in Fig. 3.9. There is no net rotation because the surface is intrinsically flat and has $K = 0$.

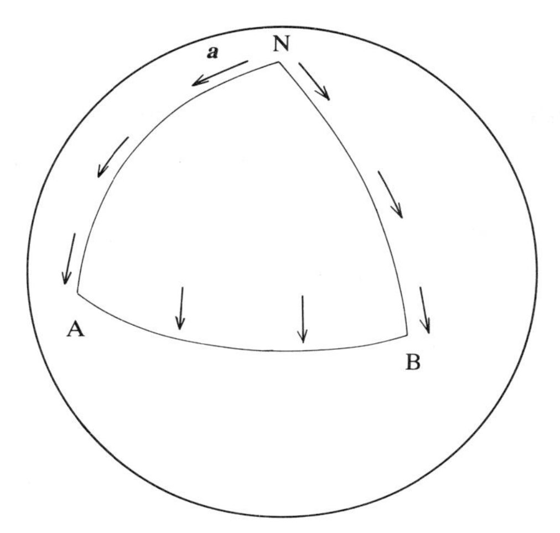

Fig. 3.8 The parallel transport of a vector around a closed path NABN over a spherical surface.

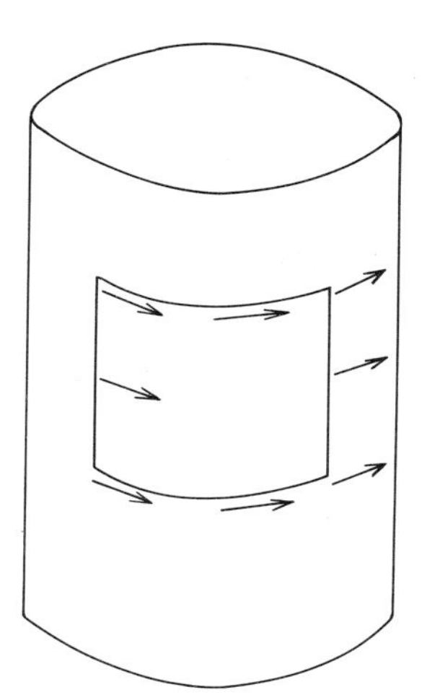

Fig. 3.9 The parallel transport of a vector around a closed path over a cylindrical surface.

3.4 Curvature and the metric equation

It has been remarked that the metric equation contains the full information on the geometry of that space. One consequence of the primacy of the metric equation is that the Gaussian curvature is a function of the coefficients g_{11}, g_{12}, and g_{22} in the metric equation. This relationship was discovered by Gauss who recognized its importance by referring to it as the "Excellent' Theorem.

For the present we shall derive the relation using an azimuthally symmetric two-dimensional surface. Later, in Chapter 7 and Appendix B, the theorem will be presented in tensor form. The space has a metric equation

$$ds^2 = g_{rr}\,dr^2 + r^2\,d\theta^2$$

where g_{rr} is a function of r only. Figure 3.10 shows a radial section through this surface embedded in a three-dimensional space, with the additional dimension lying in the paper perpendicular to r. The tangential dimension is perpendicular to the paper. The distance along the curve is

$$ds = g_{rr}^{1/2}\,dr.$$

In the figure the curvature in the section shown is

$$\frac{1}{\rho_1} = K_1 = \frac{\partial\psi}{\partial s},$$

whilst the orthogonal line in the surface (a circle around the axis of symmetry) has curvature

$$C = 1/r.$$

The surface normal lies along the line PC so that

$$K_2 = C\sin\psi = \frac{\sin\psi}{r}.$$

Then from the definition of Gaussian curvature

$$K = K_1 K_2 = \frac{\partial\psi}{\partial s}\frac{\sin\psi}{r}.$$

Now

$$\cos\psi = \frac{\partial r}{\partial s} = \frac{1}{g_{rr}^{1/2}}$$

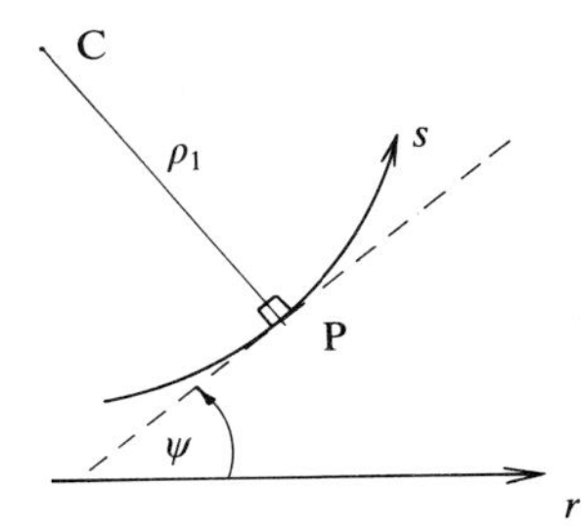

Fig. 3.10 A radial section through a two-dimensional cylindrically symmetric curved surface; r is the coordinate distance from the centre and s is the distance measured over the surface from the centre.

and therefore

$$\sin^2 \psi = 1 - 1/g_{rr}.$$

Differentiating this last expression with respect to s gives

$$2 \sin \psi \cos \psi \frac{\partial \psi}{\partial s} = \frac{(\partial g_{rr}/\partial r)(\partial r/\partial s)}{g_{rr}^2}.$$

Therefore

$$K = \frac{\partial g_{rr}/\partial r}{2rg_{rr}^2}. \tag{3.4}$$

Spaces with more than two dimensions require more than a single parameter to describe the Gaussian curvature at a given point. The exact count of curvature parameters analogous to K is easy enough to make. First a pair of geodesics **a** and **b** are drawn through a point in the space. From these a geodesic surface can be formed which contains all the linear combinations $\lambda\mathbf{a} + \mu\mathbf{b}$, all of which are geodesics. In the case of a two-dimensional spherical surface it would simply reproduce the sphere. Such a surface has a definite Gaussian curvature $K_{\mathbf{ab}}$. Next a geodesic **c** orthogonal to **a** and **b** is selected. From **a** and **c** a second geodesic surface can be constructed whose Gaussian curvature is $K_{\mathbf{ac}}$ and is entirely independent of $K_{\mathbf{ab}}$. Continuing in this way we find $n(n-1)/2$ such independent Gaussian curvatures for a space of n dimensions. A more complete and compact description of curvature in n dimensions is embodied in the Riemann tensor. This is defined in Chapter 7 and its relationship to the Gaussian curvature is also discussed there.

3.5 Three-dimensional spaces

A simple example, and one of interest because it introduces ideas of value in thinking about cosmological models, is a space which is both homogeneous and isotropic. In an isotropic space all geodesic surfaces at a single point have by definition identical Gaussian curvature: in a space that is homogeneous as well the curvature is the same everywhere. Polar coordinates (r, θ, φ) are indicated, where θ is the polar angle and φ is the azimuthal angle. (With the origin at the centre of the Earth θ would run from 0° at the North Pole to 180° at the South Pole, while φ (the longitude) runs from $-180°$ to $+180°$.) Small changes in r, θ, or φ individually give movements that are mutually orthogonal. First consider displacements in a flat three-dimensional space, which is one sort of isotropic homogeneous space. A change of dr in r moves a radial distance dr, a change dθ moves a distance $r\,\mathrm{d}\theta$ along a line of longitude, and a change of dφ moves a distance $r \sin\theta\, \mathrm{d}\varphi$ along a line of latitude ($r \sin \theta$ is the radius of the line of latitude). The overall interval moved is

$$\mathrm{d}s^2 = \mathrm{d}r^2 + r^2\, \mathrm{d}\theta^2 + r^2 \sin^2\theta\, \mathrm{d}\varphi^2 \tag{3.5}$$

which is the metric equation of flat space in polar coordinates. In shorthand this can be expressed as

$$ds^2 = dr^2 + r^2\, d\Omega^2.$$

Next consider a curved space. A sphere in this space centred on the origin and passing through the coordinate r will have an area $4\pi r^2$ and a circumference $2\pi r$. However, the distance from the origin to the sphere is no longer r but rather $\int g_{rr}^{1/2}\, dr$ where g_{rr} changes with the distance r. Thus the metric equation is

$$ds^2 = g_{rr}\, dr^2 + r^2\, d\Omega^2.$$

Let us now take an equatorial section through this space. θ is $\pi/2$ in this section and

$$ds^2 = g_{rr}\, dr^2 + r^2\, d\varphi^2.$$

This surface has the metric equation that we met earlier in Section 3.4. There the curvature of this surface was shown to be

$$K = \frac{dg_{rr}/dr}{2rg_{rr}^2}.$$

Rearrangement gives

$$\frac{dg_{rr}}{g_{rr}^2} = 2Kr\, dr.$$

At this point we make use of the homogeneity and isotropy of the space. K will have the same value for all geodesic surfaces and at all points. This makes the integration simple; it gives

$$1/g_{rr} = C - Kr^2$$

where C is the constant of integration. C can be determined by considering the limiting case of a flat space for which g_{rr} is unity and K is zero; hence in turn C must be unity. Thus

$$g_{rr} = \frac{1}{1 - Kr^2}.$$

We are now in a position to calculate the radial path length to the surface labelled by r:

$$s = \int_0^r \frac{dr}{(1 - Kr^2)^{1/2}}$$
$$= \frac{\sin^{-1}(rK^{1/2})}{K^{1/2}}$$

for the case when K is positive. In terms of s the area of the spherical surface (at coordinate r) is

$$A = \frac{4\pi}{K} \sin^2(sK^{1/2}). \tag{3.6}$$

An observer moving out from the origin would find that initially the area of the spherical surface he reached would grow steadily with s until he had travelled a distance $\pi/2K^{1/2}$. Beyond this point he would observe something very strange: the area of the spherical surface would decrease, until when he had travelled twice as far to $\pi/K^{1/2}$ he would find that it had shrunk to nothing. This paradox can be understood by reference to the analogous two-dimensional situation. A two-dimensional observer moving out across the surface of a sphere starting from the north pole would note that the lines of latitude get longer and longer as he approaches the equator. Once into the southern hemisphere the lines of latitude would become shorter and finally vanish at the south pole. Beyond this point the observer would be returning to the north pole. In a like manner the traveller in a three-dimensional, homogeneous, isotropic, positively curved space would eventually return to his starting point. Such spaces with positive curvature are therefore *closed spaces* of finite size.

The situation is very different in a homogeneous, isotropic, curved space with K negative. Then the expression for the area of the spherical surface labelled by r is

$$A = \frac{4\pi}{|K|} \sinh^2(s|K|^{1/2}) \tag{3.7}$$

which expands indefinitely as the distance s from the origin increases. A traveller who moves out radially from his origin will never return to his starting point. Spaces with negative curvature are therefore *open spaces.*

The reader should note that, although a definite point in the space was chosen as origin, this point could have been selected anywhere. It is the essence of a homogeneous space that the space should look the same from any point.

4

Space–time curvature

The analysis carried out in Chapter 3 for curved spaces will be extended here to the curved space–time that we inhabit. This extension is straightforward because the curved space–time we inhabit bears the same relation to the Minkowski space of SR that a curved space does to the Euclidean space of Newtonian mechanics. One situation is of particular interest: the empty space–time around a spherically symmetric mass distribution. It was the first case for which the exact solution of Einstein's equation was obtained, and this was done by Schwarzschild in 1916. The practical importance of this solution is that it is adequate for use in calculations of the general relativistic effects observed in the solar system, and of the properties of the simplest type of black hole. Section 4.1 is used to introduce the metric equation of space–time. In Section 4.2 a physical interpretation is provided for a geodesic in space–time, namely the path of a body in free fall. Then the curvature of geodesic surfaces in space–time around a spherically symmetric mass distribution are calculated or inferred. These results are combined with information from Chapter 2 on the distortion of time by the Earth's gravitational field to yield the Schwarzschild solution (Section 4.3). This simplistic approach gives a first conceptual view of the Schwarzschild solution, free of mathematical detail. Later, in Appendix D, the Schwarzschild metric will be obtained directly by solving Einstein's equation of general relativity.

4.1 The metric equation

The space–time of SR is flat and one Cartesian coordinate system (ct, x, y, z) can be used throughout all space–time. Intervals of proper time between events with coordinate separation $(c\,\mathrm{d}t, \mathrm{d}x, \mathrm{d}y, \mathrm{d}z)$ are given by

$$\mathrm{d}s^2 = c^2\,\mathrm{d}\tau^2 = c^2\,\mathrm{d}t^2 - \mathrm{d}x^2 - \mathrm{d}y^2 - \mathrm{d}z^2,$$

which is the metric equation for this space–time (Minkowski space). This metric equation differs from a metric equation valid for Euclidean space in having both positive and negative terms on the right-hand side. The set of coefficients on the right-hand side (in this case $+1, -1, -1, -1$) is called the signature of the space and labels the properties of the space. A space with some coefficients negative and some positive is called a pseudo-Euclidean space. The coefficients can always be reduced to $+1$ or -1 by appropriate choices of the length and time units. With the notation of Chapter 1 the metric equation can be rewritten

$$ds^2 = (dx^0)^2 - (dx^1)^2 - (dx^2)^2 - (dx^3)^2,$$

where $x^0 = ct$, $x^1 = x$, $x^2 = y$, and $x^3 = z$. A more compact way of writing this is

$$ds^2 = \sum_{\mu=0}^{3} \sum_{\nu=0}^{3} \eta_{\mu\nu}\, dx^\mu\, dx^\nu, \tag{4.1}$$

where the components $\eta_{\mu\nu}$ form a square matrix:

$$\eta_{\mu\nu} = \begin{matrix} 1 & 0 & 0 & 0 \\ 0 & -1 & 0 & 0 \\ 0 & 0 & -1 & 0 \\ 0 & 0 & 0 & -1. \end{matrix}$$

$\eta_{\mu\nu}$ is called a metric tensor. Subscripts and superscripts are introduced in eqn (4.1) so that the notation is consistent with that used from Chapter 5 onwards. For the present the reader can take superscripts and subscripts to be equivalent. In cases for which the system has spherical symmetry, such as space–time around a spherical star or planet, it may often be better to use polar coordinates; then

$$\begin{aligned} ds^2 &= c^2\, dt^2 - dr^2 - r^2\, d\theta^2 - r^2 \sin^2\theta\, d\varphi^2 \\ &= \sum_{\mu=0}^{3} \sum_{\nu=0}^{3} \eta_{\mu\nu}\, dw^\mu\, dw^\nu \end{aligned} \tag{4.2}$$

where $w^0 = ct$, $w^1 = r$, $w^2 = \theta$, and $w^3 = \varphi$. This equation has a different, but still diagonal, metric tensor

$$\eta_{\mu\nu} = \begin{matrix} 1 & 0 & 0 & 0 \\ 0 & -1 & 0 & 0 \\ 0 & 0 & -r^2 & 0 \\ 0 & 0 & 0 & -r^2 \sin^2\theta. \end{matrix}$$

The notation can be greatly simplified by adopting the Einstein summation convention in which we sum over repeated indices. With this convention the metric equation of Minkowski space–time simplifies to

$$ds^2 = \eta_{\mu\nu}\, dx^\mu\, dx^\nu. \tag{4.3}$$

It is useful to distinguish between Roman and Greek suffixes. When Greek letters are used, as above, the summation is to be made over time and space coordinates. With this interpretation the right-hand side of eqn (4.3) is identical with the right-hand side of eqn (4.1). However if Roman letters are used the summation is only made over space coordinates (1, 2, 3). The metric equation for three-dimensional Euclidean space with Cartesian coordinates is then

$$ds^2 = a_{ij}\, dx^i\, dx^j$$

where

$$a_{ij} = \begin{matrix} 1 & 0 & 0 \\ 0 & 1 & 0 \\ 0 & 0 & 1. \end{matrix}$$

The observations of gravitational red shift and the deviation of electromagnetic waves passing near the Sun shows that real space–time is curved. Therefore flat Minkowski space provides an inadequate description and our analysis of curved space–time will need to proceed along the lines mapped out in Chapter 3 for analysing curved space. Gaussian (generalized) coordinates x^μ can be used to cover curved space–time. The proper time interval ds between events in space–time with coordinate separation dx^μ is then given by the quadratic metric equation

$$ds^2 = c^2\, d\tau^2 = g_{\mu\nu}\, dx^\mu\, dx^\nu \tag{4.4}$$

where the components of the metric tensor $g_{\mu\nu}$ are functions of the position and time. At this point the strong equivalence principle supplies a key ingredient to understanding curved space–time. This principle requires that on transforming to a freely falling frame all local experimental measurements give results in accord with SR. What this means in geometric terms is that a Minkowski frame matches the structure of real space–time locally, but not globally. In the same way as discussed in section 3.1 that a plane can be found which is tangential to a curved two dimensional surface, a Minkowski space can always be found to match space–time locally. That is to say we can select any event in space–time and always be able to find some frame for which

$$g_{\mu\nu}(y) = \eta_{\mu\nu}$$

and

$$\left.\frac{\partial g_{\mu\nu}}{\partial x^\rho}\right|_y = 0.$$

This frame is in free fall and so our space–time belongs to the category of spaces which have a quadratic metric equation and are locally flat. Such spaces, known as pseudo-Riemann spaces, were introduced in Chapter 3.

Riemann, who was a student of Gauss, initiated the analysis of curved spaces with more than two dimensions in 1846. By the early years of the present century the mathematical properties of Riemann spaces had been extensively studied and this material was available for Einstein to use. Einstein was fortunate in having a friend Marcel Grossmann who introduced him to Riemannian geometry and who worked with him until Einstein left Zurich in 1914. The formal development of GR will be outlined in Chapters 5, 6, and 7 while in the remainder of this chapter a simpler approach will be used to infer one particularly important solution of Einstein's equation. This is the solution for empty space outside a spherically symmetric mass distribution.

4.2 Geodesics, tidal acceleration and curvature

One of the ways discussed for determining the curvature of a two-dimensional surface is to measure the deviation between nearby geodesics; in a higher-dimensional space the measurement would yield the Gaussian curvature of the two-dimensional surface containing these geodesics. This technique is equally valid in curved space–time, provided that we can first identify what a geodesic in space–time might be. A geodesic in flat space (a straight line) is the path of a free body as described by Newton's first law of motion. Equally, a time-like geodesic in Minkowski space–time (also a straight line) is the path of a free body. The equivalence principle indicates a way of transferring the definition of a geodesic to curved space–time. If a time-like geodesic in Minkowski space is the path of a free body then, to be consistent, a time-like geodesic in curved space–time should be the path of a body in free fall. This then constitutes the physical interpretation of a geodesic in curved space–time, namely the path of a body in free fall.

Having identified geodesics in space–time, let us consider the geodesic deviation between the paths of two nearby test bodies in free fall towards a spherically symmetric star. Figure 4.1 shows two such bodies A and B, both having mass m and both at a radial distance r from the star of mass M; A and B are a tangential distance ξ apart. If m is sufficiently small we can ignore the mutual attraction of the two masses. Resolving the gravitational force due to the star tangential to OA gives zero for A, while for B the force has a component

$$F_t = \frac{GMm}{r^2}\frac{\xi}{r}$$

towards A. Therefore there is a relative acceleration between A and B given by

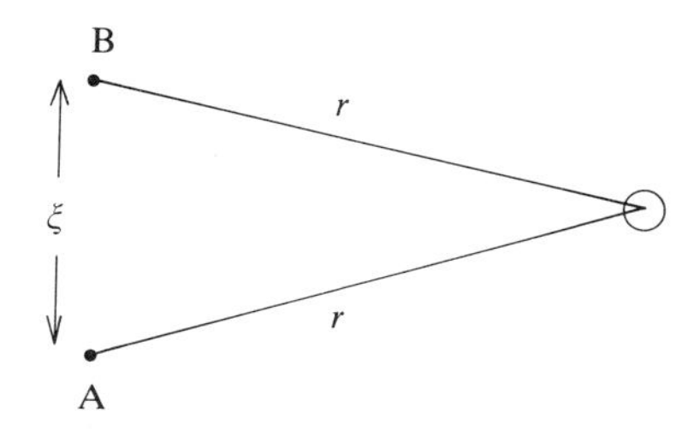

Fig. 4.1 The paths of two nearby masses A and B in radial free fall towards a star whose centre of mass lies at O. ξ is their tangential separation.

$$\frac{\mathrm{d}^2\xi}{\mathrm{d}t^2} = \frac{-GM\xi}{r^3} \tag{4.5}$$

which is independent of the masses of A and B. Such an acceleration is called a differential or *tidal* acceleration; it is unaffected by the choice of reference frame. Transforming to the frame in free fall of B, for example, does not remove the relative acceleration between A and B. Tidal accelerations are non-local and are characteristic of a curved space–time. They could be avoided only if the gravitational acceleration were the same everywhere. Then a single transformation to a frame with this acceleration would enable us to impose a frame in free fall everywhere; this is the situation described by SR. Let us take the motion of A and B to be non-relativistic, so that $\mathrm{d}x^i \ll c\,\mathrm{d}t$, and $\mathrm{d}s^2 = c^2\,\mathrm{d}\tau^2 \approx c^2\,\mathrm{d}t^2$. Then the tidal acceleration can be written

$$\frac{\mathrm{d}^2\xi}{\mathrm{d}s^2} = -\frac{GM}{r^3c^2}\,\xi.$$

This expression can be compared with eqn (3.2) for the geodesic deviation over a curved surface. However, we must take account of the sign change for the spatial components of the metric equation on going from three-space to space–time, and so

$$\frac{\mathrm{d}^2\eta}{\mathrm{d}s^2} = K\eta \tag{4.6}$$

where K is the Gaussian curvature. The tidal acceleration is thus related to space–time curvature. In the case considered the motion is transverse to OA, and so the curvature is that of the geodesic surface defined by the time direction and the spatial direction transverse to OA. Without any loss of generality we can take AOB to be in the equatorial plane of the parent star so that the acceleration is purely in the coordinate φ. From the comparison between the preceding two equations, the Gaussian curvature of the geodesic surface (φ, t) is deduced to be

$$K_{\varphi t} = \frac{-GM}{r^3c^2}. \tag{4.7}$$

4.3 The Schwarzschild metric

The analysis of the gravitational red shift led to the relation (eqn 2.2)

$$\mathrm{d}\tau^2 = \mathrm{d}t^2\left(1 - \frac{2GM}{rc^2}\right).$$

From the present viewpoint this can be regarded as the time component of the metric equation for space–time round a spherically symmetric mass. What is needed in order to complete the metric equation is information about the spatial terms. A first step would be to determine the curvature of a purely spatial surface and then apply eqn (3.4) to obtain g_{rr}. Unfortunately the method described in the last section for determining curvature fails in the case of purely spatial surfaces. Their curvatures can only be obtained rigorously from the solution of Einstein's equation. For the present we need to proceed by inferences. Firstly, the curvatures are likely to depend linearly on mass; otherwise a single mass $M + M'$ would produce a different effect from that produced by two separate masses M and M' brought into contact. Secondly, the only other relevant quantities on which the curvature might depend are G, r, and c. Dimensional analysis then shows that the simplest combination of these quantities which has the dimensions of curvature is GM/r^3c^2, and indeed the curvature $K_{\varphi t}$ has just this form. An inference is that the curvature of the spatial equatorial surface $r - \varphi$ with $\theta = \pi/2$ is

$$K_{r\varphi} = \frac{-GM}{r^3c^2}, \tag{4.8}$$

where the negative sign has been selected so as to reproduce the experimentally observed deviation of light. At this point we need to make use of another result from the preceding chapter, namely that the Gaussian curvature of an equatorial surface is related to the radial component of the metric tensor. From eqn (3.4) we have

$$K = \frac{\mathrm{d}g_{rr}/\mathrm{d}r}{2g_{rr}^2 r}.$$

Substituting the value inferred for $K_{r\varphi}$ into this expression gives

$$\frac{\mathrm{d}g_{rr}}{g_{rr}^2} = -\frac{2GM}{r^2c^2}\,\mathrm{d}r.$$

Integrating this equation gives

$$\left[\frac{-1}{g_{rr}}\right] = \left[\frac{2GM}{rc^2}\right],$$

and taking limits of integration to be $r = r$ and $r = \infty$ gives

$$g_{rr} = \frac{1}{1 - 2GM/rc^2}.$$

This suggests that the spatial separation of nearby points in a spherically symmetric space is

$$\mathrm{d}s^2 = \frac{\mathrm{d}r^2}{1 - 2GM/rc^2} + r^2\,\mathrm{d}\theta^2 + r^2\sin^2\theta\,\mathrm{d}\varphi^2.$$

This can be rewritten as

$$\mathrm{d}s^2 = \frac{\mathrm{d}r^2}{1 - 2GM/rc^2} + r^2\,\mathrm{d}\Omega^2 \tag{4.9}$$

with the notation $\mathrm{d}\Omega^2 = \mathrm{d}\theta^2 + \sin^2\theta\,\mathrm{d}\varphi^2$. The complementary expression for the effect of the mass distribution on time has been given in eqn (2.2). Equations (4.9) and (2.2) can now be combined to give the metric equation for empty space outside a spherically symmetric mass distribution:

$$\mathrm{d}s^2 = c^2\,\mathrm{d}\tau^2 = c^2\,\mathrm{d}t^2\left(1 - \frac{2GM}{rc^2}\right) - \frac{\mathrm{d}r^2}{1 - 2GM/rc^2} - r^2\,\mathrm{d}\Omega^2. \qquad (4.10)$$

Note that the relative signs of the spatial components with respect to the time component are set negative, so that in the limit of zero mass eqn (4.10) reduces to the Minkowski metric equation of SR:

$$\mathrm{d}s^2 = c^2\,\mathrm{d}\tau^2 = c^2\,\mathrm{d}t^2 - \mathrm{d}r^2 - r^2\,\mathrm{d}\Omega^2.$$

Equation (4.10) is known as the Schwarzschild metric equation for which the metric tensor is

$$g_{\mu\nu} = \begin{pmatrix} \left(1 - \dfrac{2GM}{rc^2}\right) & & & \\ & -\left(1 - \dfrac{2GM}{rc^2}\right)^{-1} & & \\ & & -r^2 & \\ & & & -r^2\sin^2\theta \end{pmatrix}. \qquad (4.11)$$

Note that the components of eqn (4.11) do not depend on time, and so the Schwarzschild metric is static. Birkhoff proved that this metric equation is the unique description in GR for space–time outside a spherically symmetric mass distribution (carrying no charge or angular momentum). It follows that even when the mass is undergoing spherically symmetric motion the external structure of space–time is described by the Schwarzschild metric and hence is static. This result would apply in the case that a stellar core collapsed to a neutron star or to a black hole while maintaining perfect spherical symmetry. The Schwarzschild metric will be used in calculations of the observable effects of GR in the solar system (Chapter 8) and of the properties of non-rotating neutral black holes (Chapter 9).

It is important to retain a firm grasp on what the coordinates appearing in eqn (4.11) are and what they are not. Referring back to eqn (2.2) we recall that $\mathrm{d}t$ is the interval of time between two events measured by an observer using a clock which is in a region remote enough that space–time is effectively flat; $\mathrm{d}t$ is called the coordinate time interval. However, $\mathrm{d}\tau$ is the proper time interval measured on a clock carried by someone moving from (t, r, θ, φ) to $(t + \mathrm{d}t, r + \mathrm{d}r, \theta + \mathrm{d}\theta, \varphi + \mathrm{d}\varphi)$. The coordinate r is different from the radial distance measured from the centre of the mass M, which we shall call a. The relationship between a and r is

$$\mathrm{d}a = \left(\frac{1}{1 - 2GM/rc^2}\right)^{1/2} \mathrm{d}r.$$

The area of the spherical surface labelled by r is $4\pi r^2$ and not $4\pi a^2$, and its circumference is $2\pi r$ and not $2\pi a$.

From the form of the Schwarzschild metric equation it is clear that the factor $2GM/rc^2$ is an important measure of the effect of mass on the curvature of space–time. When $2GM/rc^2$ is small compared with unity, the curvature is small and general relativistic effects are negligible. Conversely, if $2GM/rc^2$ approaches unity the curvature is severe and general relativistic effects dominate. Space–time in the vicinity of a star whose radius shrinks to a value less than $r_0 = 2GM/c^2$ becomes so warped that the region inside radius r_0 is effectively isolated from the rest of the Universe. This is the phenomenon called a black hole and is the subject of Chapter 9. The metric components can also be rewritten in terms of the gravitational potential $\varphi = -GM/r$ so that $g_{00} = -1/g_{11} = 1 + 2\varphi/c^2$. Then if gravitational forces are interpreted as the consequence of space–time curvature there is a correspondence between the single gravitational potential of Newtonian mechanics on the one hand and the metric components on the other hand.

5
Elementary tensor analysis

In the preceding chapters a simple presentation has been made of the properties of curved space–time. Now the general theory of relativity will be developed formally using tensors. Equations between tensors have the property that they remain valid under the general transformations between Gaussian coordinate systems in curved space–time. Therefore physical laws when expressed in terms of tensors will retain their form under these general transformations, in particular under changes to accelerating frames. It is this property which is relevant for building GR. Those elements of tensor analysis which are required for assimilating GR are presented in this chapter.

Scalars and vectors are in fact the simplest types of tensors, and from SR it is well known that if an equality can be proved between vectors (or scalars) in one inertial frame then the equality remains true under Lorentz transformations to other inertial frames. The new feature for curved space–time is that the permitted transformations are quite general and not simply linear, as in the case with Lorentz transformations. For example the acceleration in a constant uniform gravitational field yields a quadratic relation between position and time of the form

$$x' = x - gt^2/2,$$

while acceleration in realistic gravitational fields which depend on position and time will give more complicated transformations. Space–time derivatives, such as momentum, which behave as vectors under Lorentz transformations need redefinition in order to behave as vectors under general transformations. The method for recasting all space–time derivatives so that they are all tensors under general coordinate transformations is described in Chapter 6. At that stage a simple procedure can be defined for writing physical laws in a form compatible with GR. Take a physical law valid in SR, which is thus true in a frame in a free fall, and recast this in tensor form. The resulting tensor equation is automatically valid under all general transformations. This technique fails for the central question of how to write the relativistic law of gravitation because Newton's law of gravitation is not compatible with SR and so we lack a starting point. Put another way, SR is only valid locally in a frame in free fall, which neatly disconnects it from any gravitational effects. What Einstein did was to suggest that there exists a tensor identity between space–time curvature and a tensor, the stress-energy tensor, describing the distribution of matter. This identity provides the relativistic law of gravitation and forms the keystone of GR. The curvature and stress-energy tensors, and Einstein's equation that links them, will be discussed in Chapter 7.

5.1 General transformations

General transformations in curved space–time are non-linear. Suppose that one set of Gaussian coordinates which covers space–time is $(x^0, x^1, x^2, x^3) = x^\mu$, and a second set is x'^μ. The x'^μs are more or less complicated functions of the x^μs:

$$x'^\mu = x'^\mu(x^0, x^1, x^2, x^3).$$

Equally the x^μs are functions of the x'^μs:

$$x^\mu = x^\mu(x'^0, x'^1, x'^2, x'^3).$$

These relations between *finite* coordinate distances are generally so inconvenient that it makes more sense to start calculations from the differentials which do transform linearly:

$$\mathrm{d}x'^\mu = \frac{\partial x'^\mu}{\partial x^\nu}\,\mathrm{d}x^\nu. \tag{5.1}$$

Recall that the summation convention requires the right-hand side of this equation to be interpreted as

$$\frac{\partial x'^\mu}{\partial x^0}\,\mathrm{d}x^0 + \frac{\partial x'^\mu}{\partial x^1}\,\mathrm{d}x^1 + \frac{\partial x'^\mu}{\partial x^2}\,\mathrm{d}x^2 + \frac{\partial x'^\mu}{\partial x^3}\,\mathrm{d}x^3.$$

Another notational simplification is to define

$$\Lambda^\mu{}_\nu = \frac{\partial x'^\mu}{\partial x^\nu} \qquad \text{and} \qquad \Lambda_\mu{}^\nu = \frac{\partial x^\nu}{\partial x'^\mu}.$$

In these definitions the order of the superscript and subscript is significant, because in general $\Lambda^\mu{}_\nu \neq \Lambda_\nu{}^\mu$. The product

$$\Lambda^\alpha{}_\nu \Lambda_\beta{}^\nu = \frac{\partial x'^\alpha}{\partial x^\nu}\frac{\partial x^\nu}{\partial x'^\beta} = \frac{\partial x'^\alpha}{\partial x'^\beta} = \begin{cases} 1 \text{ if } \alpha = \beta \\ 0 \text{ if } \alpha \neq \beta \end{cases}.$$

This can be expressed compactly as

$$\Lambda^\alpha{}_\nu \Lambda_\beta{}^\nu = \delta^\alpha{}_\beta \tag{5.2}$$

where the Kronecker delta $\delta^\alpha{}_\beta$ is defined to be unity if α is equal to β but zero otherwise.

The local vectors discussed in Section 3.3 can be expressed in terms of a set of basis vectors $\boldsymbol{e}_\mu$ drawn along the local space–time coordinate directions at an event P. Here the subscript μ on $\boldsymbol{e}_\mu$ indicates that this basis vector points in the μ direction. For example if $\boldsymbol{e}_1$ is of unit length its components would be 0, 1, 0, and 0 along the ct, x, y, and z coordinate directions respectively; we would write these components $(\boldsymbol{e}_1)^\alpha$. If P′ is separated from P by an infinitesimal distance $\mathrm{d}x^\mu$ then PP′ is the local vector

$$\mathrm{d}x^\mu \boldsymbol{e}_\mu.$$

The invariant length of PP′ is

$$\mathrm{d}s^2 = (\mathrm{d}x^\mu \boldsymbol{e}_\mu)\cdot(\mathrm{d}x^\nu \boldsymbol{e}_\nu) = \boldsymbol{e}_\mu \cdot \boldsymbol{e}_\nu \, \mathrm{d}x^\mu \, \mathrm{d}x^\nu,$$

where $\boldsymbol{e}_\mu \cdot \boldsymbol{e}_\nu$ is a more general form of a scalar product. In Minkowski space–time with orthogonal coordinates ct, x, y, and z, by comparison with eqn (4.1), we have

$$\boldsymbol{e}_\mu \cdot \boldsymbol{e}_\nu = \eta_{\mu\nu}.$$

More generally by comparison with eqn (4.4) we have

$$g_{\mu\nu} = \boldsymbol{e}_\mu \cdot \boldsymbol{e}_\nu. \tag{5.3}$$

If the basis vectors are of unit length (and the signatures of the μ and ν components are the same, i.e. both $+1$ or both -1), then $g_{\mu\nu}$ is the cosine of the angle between $\boldsymbol{e}_\mu$ and $\boldsymbol{e}_\nu$.

All local four-vectors at the same event in space–time can be expressed in terms of the same set of basis vectors. Choosing the four-momentum, it can be written

$$p^\mu \boldsymbol{e}_\mu.$$

The components of the momentum (or any other local four-vector at the same event) will have to behave under general transformations just like the components $\mathrm{d}x^\mu$; using eqn (5.1)

$$p'^\mu = \frac{\partial x'^\mu}{\partial x^\nu} p^\nu.$$

This result has implications for the invariant length of a four-vector. In the case of four-momentum this is given in a frame in free fall where SR is valid by

$$m^2 c^2 = \eta_{\mu\nu} p^\mu p^\nu$$

where m is the system's rest mass. The expression for $\mathrm{d}s^2$ in the same frame is

$$\mathrm{d}s^2 = \eta_{\mu\nu} \, \mathrm{d}x^\mu \, \mathrm{d}x^\nu.$$

Under a general transformation $\mathrm{d}s^2$ does not change; in the new coordinates

$$\mathrm{d}s^2 = g_{\mu\nu} \, \mathrm{d}x^\mu \, \mathrm{d}x^\nu.$$

Thus, because the components of all four-vectors transform in precisely the same way, it follows that

$$m^2 c^2 = g_{\mu\nu} p^\mu p^\nu.$$

Therefore the invariant length of *any* local four-vector is unaffected by general transformations.

Next some examples of transformations of interest are given, using orthonormal local coordinates. The first is the rotation through θ around the z axis. Locally it has the same form as a rotation in SR:

$$\Lambda^{\mu}{}_{\nu} = \begin{matrix} 1 & & & \\ & \cos\theta & \sin\theta & \\ & \sin\theta & -\cos\theta & \\ & & & 1. \end{matrix}$$

A Lorentz transformation to a frame with velocity βc along the x axis has the form

$$\Lambda^{\mu}{}_{\nu} = \begin{matrix} \cosh u & -\sinh u & & \\ -\sinh u & \cosh u & & \\ & & 1 & \\ & & & 1, \end{matrix}$$

where $\cosh u = \gamma$, $\sinh u = \beta\gamma$ and $\gamma = 1/(1 - \beta^2)^{1/2}$. Finally, consider the transformation between frames momentarily coincident in velocity near a massive spherically symmetric body. The first is a frame in radial free fall, and the second is a frame at rest:

$$\Lambda^{\mu}{}_{\nu} = \begin{matrix} (1 + 2\varphi/c^2)^{1/2} & & & \\ & 1/(1 + 2\varphi/c^2)^{1/2} & & \\ & & 1 & \\ & & & 1 \end{matrix}$$

where the 1-axis is radial and φ is the gravitational potential.

5.2. Vector and covector components

Thus far all vectors have been written in terms of the familiar vector components, e.g. $dx^{\circ} = c\,dt$, $dx^1 = dx$, $dx^2 = dy$, and $dx^3 = dz$. Other components, called covector components, will now be introduced. For the purpose of discussion we shall consider a conservative force in classical mechanics and we shall work in space only. The components of a conservative force are given by

$$f_i = -\frac{\partial\varphi}{\partial x^i}$$

where φ is the potential, which is a scalar function of position only. At any given location the value of φ is invariant under transformations, i.e. it remains the same. Thus the effect of a transformation on the components of f_i is given by

$$f_i' = -\frac{\partial \varphi}{\partial x'^i}$$

$$= -\frac{\partial \varphi}{\partial x^j}\frac{\partial x^j}{\partial x'^i}$$

$$= \frac{\partial x^j}{\partial x'^i} f_j. \tag{5.4}$$

This is the inverse of the transformation given in eqn (5.1); f_i transforms like d/dx^i rather than like dx^i. Components which transform in this new way are called covariant components; they are written with a subscript. Components that transform in the manner of eqn (5.1) are called vector or contravariant components. The f_i are the components of the gradient of φ, and so a useful physical view is that covector components make up a *gradient*. Consider next a curve in space $x^\mu(s)$ where s measures the curve length from some reference point along it. At any location on the curve the components dx^μ/ds make up a tangent vector, and this transforms according to eqn (5.1). Therefore a physical interpretation of vector components is that they make up a *tangent*. Next we shall show that *any* vector can be expressed in terms of either its vector or covector components. Depending on the context one set or other may be of more physical importance.

The local interval PP′ used in Section 5.1 has

$$ds^2 = g_{\mu\nu}\, dx^\mu\, dx^\nu.$$

Let us write

$$dx_\mu = g_{\mu\nu}\, dx^\nu \tag{5.5}$$

and then

$$ds^2 = dx_\mu\, dx^\mu. \tag{5.6}$$

The quantity ds^2 is invariant under any general transformation; hence

$$\begin{aligned} ds^2 &= dx'_\alpha\, dx'^\alpha \\ &= dx'_\alpha \frac{\partial x'^\alpha}{\partial x^\mu}\, dx^\mu. \end{aligned} \tag{5.7}$$

Comparing eqn (5.6) with eqn (5.7) it is seen that

$$dx'_\alpha = \frac{\partial x^\mu}{\partial x'^\alpha}\, dx_\mu.$$

In other words the dx_μ are the covector components of PP′. The operation given by eqn (5.5) is a general way of generating covector components from vector components.

A two-dimensional example can be used to illustrate these results. Figure 5.1

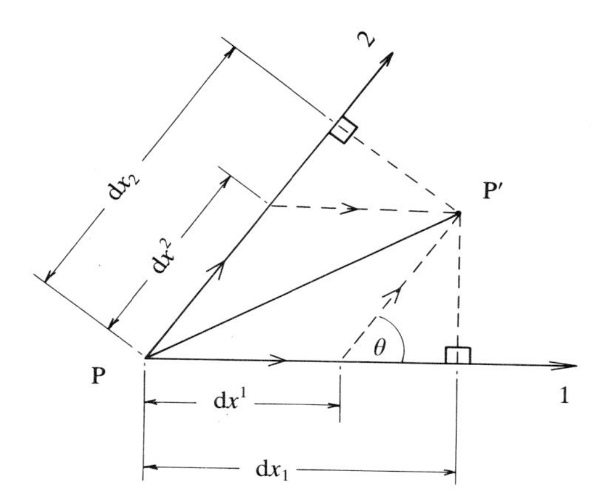

Fig. 5.1 The vector and covector components of an infinitesimal vector PP′ in a two-dimensional space.

shows a local vector PP′ in two spatial dimensions, where the local axes are inclined at an angle θ. The vector component of PP′ on the 1-axis is obtained by projecting from P′ parallel to the 2-axis, and similarly for the 2-component. Then in terms of unit vectors $\boldsymbol{e}_1$ and $\boldsymbol{e}_2$ along the axes:

$$\mathrm{PP}' = \boldsymbol{e}_1\,\mathrm{d}x^1 + \boldsymbol{e}_2\,\mathrm{d}x^2.$$

The covector components are given by eqn (5.5) for two dimensions:

$$\mathrm{d}x_1 = g_{11}\,\mathrm{d}x^1 + g_{12}\,\mathrm{d}x^2.$$

Then using eqn 5.3 we obtain

$$\mathrm{d}x_1 = \mathrm{d}x^1 + \mathrm{d}x^2 \cos\theta,$$

and similarly

$$\mathrm{d}x_2 = \mathrm{d}x^2 + \mathrm{d}x^1 \cos\theta.$$

Figure 5.1 shows that these covector components $\mathrm{d}x_1$ and $\mathrm{d}x_2$ can be obtained by projecting perpendicularly from P′ onto the relevant axis. How this generalizes in Euclidean spaces of higher dimension is illustrated in Fig. 5.2. The vector component of PP′ is obtained by projecting from P′ onto the 1-direction using the surface through P′ that contains all the other local basis vectors: $\mathrm{d}x^1$ is SP. The covector component is simply the perpendicular projection from P′ onto the 1-direction: $\mathrm{d}x_1$ is S′P. With rectangular Cartesian coordinates in Euclidean space the distinction between vector and covector components disappears, which explains why covector components are not usually used in Newtonian mechanics. Direct visualization of vector and covector components in space–time meets difficulties with the time components because we can only draw Euclidean spaces in our three-dimensional world. The covector components of local four-vectors are given by expressions

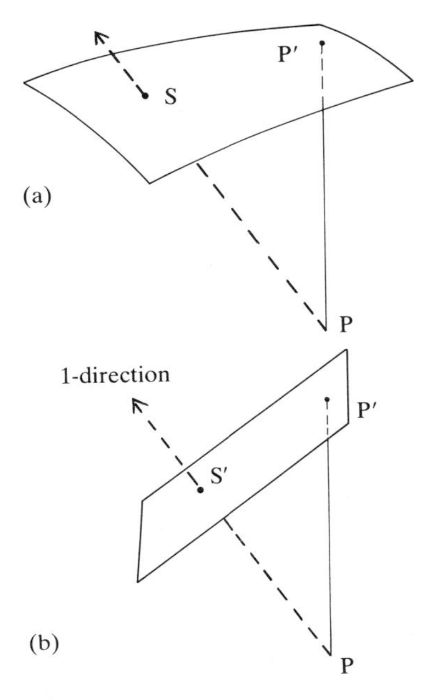

Fig. 5.2 The method for constructing (a) the vector and (b) the covector 1-component of the infinitesimal vector PP′ in a multi-dimensional space.

similar to eqn (5.5), e.g. for momentum

$$p_\mu = g_{\mu\nu} p^\nu. \tag{5.8}$$

The invariant length can therefore be rewritten as

$$m^2c^2 = p_\mu p^\mu.$$

Other related quantities, the scalar products of pairs of four-vectors, are also invariant. One example is

$$g_{\mu\nu} A^\mu B^\nu.$$

If the lowering property of $g_{\mu\nu}$ is applied in turn to A^μ and B^ν we obtain

$$g_{\mu\nu} A^\mu B^\nu = A_\nu B^\nu = A^\mu B_\mu. \tag{5.9}$$

5.3. Other tensors

The metric tensor appearing in eqn (4.5) is one example of a tensor with two indices. Another such tensor which will be of great interest is the stress-energy tensor $T^{\mu\nu}$ that summarizes the energy content of matter. Tensors having n indices are called tensors of *rank n*. Vectors are rank 1 and scalars are rank 0 tensors. The simplest examples of higher-rank tensors can be formed by taking products of components of vectors. Suppose that a^μ, b^μ, d_μ, and e_μ are vector and covector components. Then the following sets of components make up three tensors (A, B, C) of rank 2:

$$A^{\mu\nu} = a^\mu b^\nu$$

$$B^\nu{}_\mu = a^\nu d_\mu$$

$$C_{\mu\nu} = d_\mu e_\nu,$$

while

$$D^\mu_{\nu\sigma} = a^\mu d_\nu e_\sigma$$

has rank 3. General transformations do not alter the invariant interval ds^2. Thus

$$\begin{aligned} ds^2 &= g'_{\mu\nu}\, dx'^\mu\, dx'^\nu \\ &= g'_{\mu\nu} \frac{\partial x'^\mu}{\partial x^\rho} \frac{\partial x'^\nu}{\partial x^\sigma} \partial x^\sigma\, \partial x^\rho. \end{aligned}$$

In terms of the original coordinates

$$ds^2 = g_{\sigma\rho}\, dx^\sigma\, dx^\rho.$$

Comparing these last two expressions gives

$$g_{\sigma\rho} = g'{}_{\mu\nu} \frac{\partial x'^\mu}{\partial x^\rho} \frac{\partial x'^\nu}{\partial x^\sigma},$$

i.e.

$$\begin{aligned} g'_{\mu\nu} &= \frac{\partial x^\rho}{\partial x'^\mu} \frac{\partial x^\sigma}{\partial x'^\nu} g_{\sigma\rho} \\ &= \Lambda_\mu{}^\rho \Lambda_\nu{}^\sigma g_{\sigma\rho}. \end{aligned} \tag{5.10}$$

The metric tensor requires one transformation for each of its indices. This is also true for any of the tensors listed above; for example

$$\begin{aligned} D'^\mu_{\nu\sigma} &= a'^\mu d'{}_\nu e'{}_\sigma \\ &= \Lambda^\mu{}_\alpha a^\alpha \Lambda_\nu{}^\beta d_\beta \Lambda_\sigma{}^\gamma e_\gamma \\ &= \Lambda^\mu{}_\alpha \Lambda_\nu{}^\beta \Lambda_\sigma{}^\gamma D^\alpha{}_{\beta\gamma}. \end{aligned}$$

We see that each vector index brings a factor $\Lambda^*{}_*$ while each covector index brings a factor $\Lambda_*{}^*$. Valid identities between tensors only connect tensors with equal numbers of subscripts and equal numbers of superscripts (the same rank). Furthermore, if the components of two tensors can be shown to be equal in a particular coordinate system, i.e.

$$T^{\mu\nu} = G^{\mu\nu},$$

then because the corresponding components of $T^{\mu\nu}$ and $G^{\mu\nu}$ transform identically the equality holds for all coordinate frames including, of course, accelerating frames. This is the vital property that makes a tensor presentation of GR so appropriate.

There are a number of other tensor properties which are needed in Chapters 6 and 7. A frequent tensor manipulation is the process of contraction. Consider a tensor $A^{\alpha\beta}_{\gamma\beta}$ where the summation is implied for all values of the repeated index β, i.e. the tensor in full is

$$A^{\alpha 0}_{\gamma 0} + A^{\alpha 1}_{\gamma 1} + A^{\alpha 2}_{\gamma 2} + A^{\alpha 3}_{\gamma 3}.$$

Applying the usual procedure to obtain this in another frame we obtain

$$\begin{aligned} A'^{\mu\nu}_{\sigma\nu} &= \Lambda^{\mu}{}_{\alpha}\Lambda^{\nu}{}_{\beta}\Lambda_{\nu}{}^{\delta}\Lambda_{\sigma}{}^{\gamma}A^{\alpha\beta}_{\gamma\delta} \\ &= \Lambda^{\mu}{}_{\alpha}\delta^{\delta}_{\beta}\Lambda_{\sigma}{}^{\gamma}A^{\alpha\beta}_{\gamma\delta} \\ &= \Lambda^{\mu}{}_{\alpha}\Lambda_{\sigma}{}^{\gamma}A^{\alpha\beta}_{\gamma\beta} \end{aligned}$$

so that $A^{\alpha\beta}_{\gamma\beta}$ transforms as a rank 2 tensor. One more contraction gives $A^{\alpha\beta}_{\alpha\beta}$, which is a scalar quantity. The scalar products met in vector analysis are familiar examples of contraction. Other tensor products involving contraction are

$$A^{\mu\nu}B_{\nu} \qquad \text{which is a vector}$$

and

$$C^{\mu\nu\sigma}B^{\alpha}_{\sigma} \qquad \text{which is a rank 3 tensor.}$$

Not all collections of numbers or functions labelled with suffixes '$F_{\mu\nu}$' necessarily form tensors. Whether or not '$F_{\mu\nu}$' constitute a tensor can be tested using the *quotient theorem*. This theorem states that if the product of $F_{\mu\nu}$ with any arbitrary tensor is also a tensor, then $F_{\mu\nu}$ is itself a tensor. One specific case will serve to illustrate the proof of this theorem. Suppose that A^{ν} is an arbitrary vector whose contraction $F_{\mu\nu}A^{\nu}$ is a covector B_{μ}. Referred to some other coordinate choice,

$$F'_{\alpha\beta}A'^{\beta} = B'_{\alpha},$$

i.e.

$$F'_{\alpha\beta}\frac{\partial x'^{\beta}}{\partial x^{\nu}}A^{\nu} = \frac{\partial x^{\mu}}{\partial x'^{\alpha}}B_{\mu}.$$

Multiplying this result by $\partial x'^{\alpha}/\partial x^{\mu}$ gives

$$\frac{\partial x'^{\alpha}}{\partial x^{\mu}}\frac{\partial x'^{\beta}}{\partial x^{\nu}}F'_{\alpha\beta}A^{\nu} = B_{\mu}.$$

Now subtracting the original form

$$F_{\mu\nu}A^{\nu} = B_{\mu}$$

from the last equation, we obtain

$$\left(F'_{\alpha\beta}\frac{\partial x'^{\alpha}}{\partial x^{\mu}}\frac{\partial x'^{\beta}}{\partial x^{\nu}} - F_{\mu\nu}\right)A^{\nu} = 0.$$

The choice of A^{ν} is arbitrary so that the expression in parentheses must vanish identically. Thus

$$F_{\mu\nu} = F'_{\alpha\beta}\frac{\partial x'^{\alpha}}{\partial x^{\mu}}\frac{\partial x'^{\beta}}{\partial x^{\nu}}$$

and therefore $F_{\mu\nu}$ transforms like a tensor of rank 2. Similar proofs can be pictured for other rank tensors.

Finally we collect here some useful properties of the metric tensor. The metric tensor can always be chosen to be symmetric under interchange of its subscripts. Suppose for argument that $g_{\mu\nu}$ is not symmetric; then it can be decomposed into a part that is symmetric $(g_{\mu\nu} + g_{\nu\mu})/2$ and a second part that is anti-symmetric $(g_{\mu\nu} - g_{\nu\mu})/2$. The contribution to ds^2 from the anti-symmetric part is

$$\frac{(g_{\mu\nu} - g_{\nu\mu})\,dx^{\mu}\,dx^{\nu}}{2}$$

which vanishes identically, and so any anti-symmetric part of $g_{\mu\nu}$ can safely be neglected.

Contraction with $g_{\mu\nu}$ is used to generate *associated* tensors. For example,

$$A_{\nu} = g_{\nu\mu}A^{\mu} \qquad \text{and} \qquad A_{\mu\nu} = g_{\mu\alpha}g_{\nu\beta}A^{\alpha\beta}.$$

The associated tensor of $g_{\mu\nu}$ itself is also of importance. Suppose that g is the determinant of the elements of the metric tensor and that $G^{\mu\nu}$ is the cofactor of $g_{\mu\nu}$ in this determinant. Then

$$g_{\alpha\mu}G^{\mu\nu} = \delta^{\nu}_{\alpha}g.$$

Let us write

$$g^{\mu\nu} = G^{\mu\nu}/g;$$

then

$$g_{\alpha\mu}g^{\mu\nu} = \delta^{\nu}{}_{\alpha}. \tag{5.11}$$

$g^{\mu\nu}$ must be symmetric if $g_{\mu\nu}$ is symmetric. Now consider the associated tensors A_{μ} and A^{μ}:

$$A_{\alpha} = g_{\alpha\mu}A^{\mu}$$

and multiply by $g^{\beta\alpha}$. This gives

$$\begin{aligned} g^{\beta\alpha}A_{\alpha} &= g_{\alpha\mu}g^{\beta\alpha}A^{\mu} \\ &= g_{\alpha\mu}g^{\alpha\beta}A^{\mu} \\ &= \delta^{\beta}_{\mu}A^{\mu} = A^{\beta}. \end{aligned}$$

Hence the effect of $g^{\beta\alpha}$ is to *raise* a subscript while the effect of contracting a tensor with $g_{\alpha\beta}$ is to *lower* a superscript. Multiplying eqn (5.11) by $g_{\beta\nu}$ gives

$$g_{\alpha\mu} g_{\beta\nu} g^{\mu\nu} = \delta^{\nu}{}_{\alpha} g_{\beta\nu} = g_{\beta\alpha}.$$

Rearrangement yields

$$g_{\beta\alpha} = g_{\beta\nu} g_{\alpha\mu} g^{\nu\mu} \tag{5.12}$$

which shows that $g_{\beta\alpha}$ *and* $g^{\nu\mu}$ are associated tensors. In the case that $g_{\alpha\beta}$ is diagonal (e.g. for the Schwarzschild metric),

$$g_{\alpha\alpha} = g_{\alpha\nu} g_{\alpha\mu} g^{\mu\nu} = g_{\alpha\alpha} g_{\alpha\alpha} g^{\alpha\alpha}$$

with no summation implied over α in this particular case. Thus

$$g^{\alpha\alpha} = 1/g_{\alpha\alpha}, \tag{5.13}$$

again with no summation over α. Applying these results for the Schwarzschild metric when using spherical polar coordinates gives

$$\begin{aligned} g^{00} &= \frac{1}{1 - 2GM/rc^2} \\ g^{11} &= -\left(1 - \frac{2GM}{rc^2}\right) \\ g^{22} &= -1/r^2 \\ g^{33} &= -\frac{1}{r^2 \sin\theta} \end{aligned} \tag{5.14}$$

For completeness the transformations of the Kronecker delta $\delta^{\nu}{}_{\mu}$ need to be discussed. Under a general transformation

$$\begin{aligned} (\delta')^{\alpha}{}_{\beta} &= \frac{\partial x'^{\alpha}}{\partial x^{\nu}} \frac{\partial x^{\mu}}{\partial x'^{\beta}} \delta^{\nu}{}_{\mu} \\ &= \frac{\partial x'^{\alpha}}{\partial x^{\mu}} \frac{\partial x^{\mu}}{\partial x'^{\beta}} \\ &= \delta^{\alpha}{}_{\beta}, \end{aligned}$$

which demonstrates that this tensor is the same in all frames. Equation (5.11) shows that $\delta^{\alpha}{}_{\beta}$ and $g^{\alpha}{}_{\beta}$ are identical tensors:

$$g^{\alpha}{}_{\beta} = g_{\beta\mu} g^{\mu\alpha} = \delta^{\alpha}{}_{\beta}. \tag{5.15}$$

6
Einstein's theory I

The main theme of this chapter is to obtain physical laws that are valid under general transformations to an accelerated frame. According to the principle of equivalence the physical laws in any frame in free fall are consistent with SR. In Chapter 5 we discovered that physical laws expressed as tensor equations automatically retain their form under general transformations. These two ideas are now fused into the *principle of generalized covariance*, first postulated by Einstein. This states that physical laws are expressible as tensor equations which reduce to laws consistent with SR in a frame in free fall. Therefore a law valid in SR can be generalized to apply in any accelerating frame by expressing it in tensor form. Most laws are already expressed in terms of vectors, which are tensors, and so this sounds as if the laws do not need any change! However, most laws are dynamical equations containing space–time derivatives (e.g. the rate of change of momentum) which are *not* tensor quantities. A new derivative called the covariant derivative is introduced in this chapter. It is a tensor and is identical with the usual space–time derivative in a frame in free fall. Then in order to convert physical laws valid in SR (free fall) directly into forms that are valid in accelerating frames as well, it is generally sufficient to replace all space–time derivatives by their covariant equivalents.

In Section 6.1 the covariant derivative is introduced and is found to depend on entities called metric connections which quantify how local vectors change when they are transported through curved space–time. The metric connections are evaluated in Section 6.2 in terms of the derivatives of the metric coefficients $g_{\mu\nu}$. Some further features of the covariant derivative are presented in Section 6.3. Newton's second law of motion is the fundamental dynamical equation, and in Section 6.4 we rewrite it in a form valid in accelerating frames. In the absence of non-gravitational forces the second law describes a body in free fall and we show in Section 6.5 that such a path is a geodesic in space–time. A procedure for transforming to a frame in free fall is described in Section 6.6. Geodesics are also 'stationary' paths in space–time with an integral equation that will be used in later chapters when calculating the trajectories of planets and light in curved space–time (Section 6.7). Finally the correspondence between metric connections and the inertial and gravitational forces of classical (Newtonian) mechanics is explored in Section 6.8.

6.1. The covariant derivative

We consider the change in a local four-vector q^μ during some physical process. A concrete choice could be the four-momentum of a test body acted on by gravitational and non-gravitational forces. In Fig. 6.1 the broken line indicates the path followed by the body in curved space–time, where s is the path length and e_μ a basis vector in the μ-direction. Changes in the coordinate frame along the path mean that the component q^μ would change, irrespective of any physical process. Over an element Δs of path between events P and P′ the change will be written δq^μ. This must be subtracted from the observed change in the μ-component between the events P and P′ which is written Δq^μ. Thus the physically important change is

$$\Delta q^\mu - \delta q^\mu.$$

Figure 6.2 shows these vectors at P and P′ and their μ-components. A quantity called the covariant derivative can be formed by dividing this change by the path length Δs between P and P′ and taking the limit as $\Delta s \to 0$, i.e.

$$\frac{Dq^\mu}{Ds} = \mathrm{Lt}_{\Delta s \to 0} \frac{\Delta q^\mu - \delta q^\mu}{\Delta s}.$$

In the above limit the difference between the local vectors at the same point is being taken, hence the covariant derivative is a valid vector. At this point a key question remains unanswered: what is the prescription for transporting the vector q^μ at P through curved space–time to P′ so that its components at P′ and hence the changes in these components δq^μ can be evaluated? It is worth recalling that in Section 3.3 the analogous question was addressed for transport over a curved surface. There the idea of parallel transport was found to be helpful. Application of the strong equivalence principle provides the

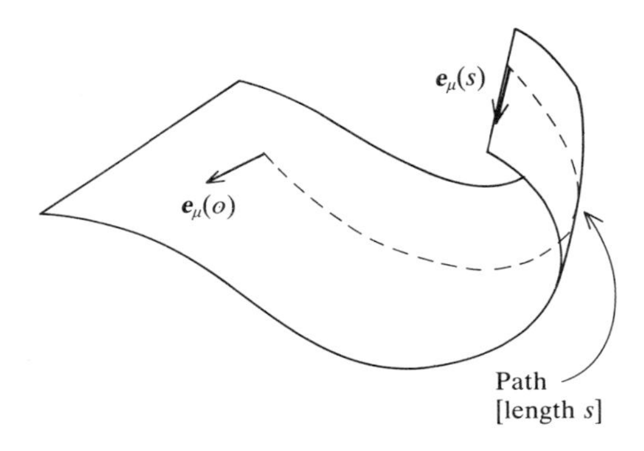

Fig. 6.1 A path across a curved surface: e_μ is a local basis vector drawn at two points along this path.

basis for a definition of parallel transport in curved space–time. The principle states that over a short enough path length it is always possible to transform to a frame in free fall; in such a frame, space–time is locally flat and the coordinates of SR may be imposed. This suggests an unambiguous procedure for transporting a local vector across curved space–time. First, the vector is transformed to a frame in free fall at P and in that frame it is carried across the interval Δs without change of its Cartesian coordinates. Then the resulting vector is transformed back to the relevant frame at P′. This procedure defines *parallel transport* of a local vector in space–time. Over a finite path a different frame in free fall may be required for each path element. Suppose the vector q^μ is parallel transported a distance Δx^ρ in the ρ-direction. Then if the vector initially had only a σ-component it would in general develop components in all directions, so

$$\delta q^\mu = -\Gamma^\mu{}_{\sigma\rho} q^\sigma \Delta x^\rho \tag{6.1}$$

This expression is linear in the displacement Δx^ρ and takes account of all the components of path (ρ), of all the initial components (σ) of the vector, and of all the components showing change (μ). The new quantities $\Gamma^\mu{}_{\sigma\rho}$ introduced

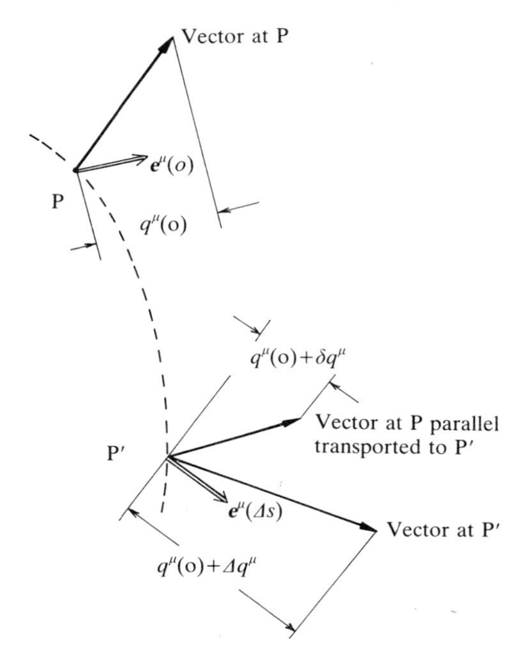

Fig. 6.2 Measurements of a vector are shown at two events P and P′ separated by a proper distance Δs. The result of parallel transporting the vector measured at P to P′ is also shown. Their components in the μ-direction are indicated. $\boldsymbol{e}_\mu$ is the basis vector marking the μ-direction.

here are called the *metric connections.* (Affine connections or affinities are more general terms used to encompass cases for which the space is not necessarily Riemannian). The metric connections are clearly functions of the position in space–time. Substituting this form for δq^μ in the definition of the covariant derivative yields

$$\frac{\mathrm{D}q^\mu}{\mathrm{D}s} = \frac{\mathrm{d}q^\mu}{\mathrm{d}s} + \Gamma^\mu{}_{\sigma\rho} q^\sigma \left(\frac{\mathrm{d}x^\rho}{\mathrm{d}s}\right). \tag{6.2}$$

$\mathrm{D}q^\mu/\mathrm{D}s$ is the μ-component of the covariant derivative of the vector with components q^μ. It is quite easy to see that the connections are not themselves tensors. In a frame in free fall the changes δq^μ will vanish and hence the connections vanish, however in other frames neither δq^μ nor the connections vanish. Therefore the connections are not tensors. On the right hand side of eqn (6.2) both terms alter under a general transformation but their sum is always the component of the same local vector

$$\frac{\mathrm{D}q^\mu}{\mathrm{D}s}.$$

Expressions for covariant derivatives of tensors of any rank can be obtained, all of which are also tensors. For covector components (see problem 6.1)

$$\frac{\mathrm{D}q_\mu}{\mathrm{D}s} = \frac{\mathrm{d}q_\mu}{\mathrm{d}s} - \Gamma^\nu{}_{\mu\rho} q_\nu \frac{\mathrm{d}x^\rho}{\mathrm{d}s}.$$

In order to evaluate the covariant derivative of a tensor of rank-2 $A_{\mu\nu}$ we consider the invariant

$$\mathrm{A}_{\mu\nu} q^\mu q^\nu.$$

When $\mathrm{A}_{\mu\nu}$ and q^μ are parallel transported in a frame in free fall this invariant will not change; further it is invariant under general transformations, and will remain invariant under parallel transport in all frames. Thus

$$\delta(\mathrm{A}_{\mu\nu} q^\mu q^\nu) = 0.$$

Expanding the left-hand side this becomes

$$\delta \mathrm{A}_{\mu\nu} q^\mu q^\nu + \mathrm{A}_{\mu\nu} \delta q^\mu q^\nu + \mathrm{A}_{\mu\nu} q^\mu \delta q^\nu = 0.$$

Inserting in this formula the expression given for δq^σ in eqn 6.1 we obtain

$$\begin{aligned}\delta \mathrm{A}_{\mu\nu} q^\mu q^\nu &= \mathrm{A}_{\mu\nu}(\Gamma^\mu{}_{\tau\rho} q^\nu q^\tau + \Gamma^\nu{}_{\tau\rho} q^\tau q^\mu)\Delta x^\rho \\ &= (\mathrm{A}_{\tau\nu}\Gamma^\tau{}_{\mu\rho} q^\nu q^\mu + \mathrm{A}_{\mu\tau}\Gamma^\tau{}_{\nu\rho} q^\nu q^\mu)\Delta x^\rho.\end{aligned}$$

Then cancelling $q^\mu q^\nu$ from both sides

$$\delta \mathrm{A}_{\mu\nu} = (\Gamma^\tau{}_{\mu\rho} \mathrm{A}_{\tau\nu} + \Gamma^\tau{}_{\nu\rho} \mathrm{A}_{\mu\tau})\Delta x^\rho.$$

Consequently the covariant derivative of $\mathrm{A}_{\mu\nu}$

$$\frac{DA_{\mu\nu}}{Ds} = Lt_{\Delta s \to 0} \frac{\Delta A_{\mu\nu} - \delta A_{\mu\nu}}{\Delta s}$$

$$= \frac{dA_{\mu\nu}}{ds} - \Gamma^{\tau}{}_{\mu\rho} A_{\tau\nu} \frac{dx^{\rho}}{ds} - \Gamma^{\tau}{}_{\nu\rho} A_{\mu\tau} \frac{dx^{\rho}}{ds}.$$

6.2. The calculation of the metric connection

The full information about the structure of space–time is embodied in the metric equation, and so it is reasonable to expect that the connections should be functions of the metric coefficients $g_{\mu\nu}$. Replacing $A_{\mu\nu}$ in the previous equation by $g_{\mu\nu}$ and multiplying throughout by ds/dx^{ρ} gives

$$\frac{Dg_{\mu\nu}}{Dx^{\rho}} = \frac{\partial g_{\mu\nu}}{\partial x^{\rho}} - \Gamma^{\tau}{}_{\mu\rho} g_{\tau\nu} - \Gamma^{\tau}{}_{\nu\rho} g_{\mu\tau}.$$

Now let us specialize to the frame in free fall. There the connections vanish so that

$$\frac{Dg_{\mu\nu}}{Dx^{\rho}} = \frac{\partial g_{\mu\nu}}{\partial x^{\rho}}.$$

In addition the metric of the frame in free fall is locally that of special relativity. As discussed in chapters 3 and 4 this condition is written

$$g_{\mu\nu} = \eta_{\mu\nu} \qquad \text{and} \qquad \frac{\partial g_{\mu\nu}}{\partial x^{\rho}} = 0.$$

Then for the frame in free fall we have

$$Dg_{\mu\nu}/Dx^{\rho} = 0,$$

which, being a tensor equation, must hold in all frames. Thus in general

$$\partial g_{\mu\nu}/\partial x^{\rho} = \Gamma^{\tau}{}_{\mu\rho} g_{\tau\nu} + \Gamma^{\tau}{}_{\nu\rho} g_{\mu\tau}.$$

Now define

$$\Gamma_{\mu\nu\rho} = g_{\mu\tau} \Gamma^{\tau}_{\nu\rho}, \tag{6.3}$$

and the previous equation can be written

$$\frac{\partial g_{\mu\nu}}{\partial x^{\rho}} = \Gamma_{\nu\mu\rho} + \Gamma_{\mu\nu\rho}.$$

For convenience we introduce the *subscript comma* notation to represent a derivative:

$$g_{\mu\nu,\rho} = \frac{\partial g_{\mu\nu}}{\partial x^{\rho}}.$$

Then the previous line becomes

$$g_{\mu\nu,\rho} = \Gamma_{\nu\mu\rho} + \Gamma_{\mu\nu\rho}. \tag{6.4}$$

Equation (6.4) is inverted in Appendix C to give

$$2\Gamma_{\nu\mu\rho} = g_{\mu\nu,\rho} - g_{\rho\mu,\nu} + g_{\nu\rho,\mu}. \tag{6.5}$$

The reader can check this result by substituting for $\Gamma_{\nu\mu\rho}$ and $\Gamma_{\mu\nu\rho}$ in eqn (6.4). This expression is called the *fundamental theorem* of Riemannian geometry. It is clear that the connections are symmetric

$$\Gamma_{\sigma\rho\mu} = \Gamma_{\sigma\mu\rho}. \tag{6.6}$$

6.3. More on the covariant derivative

The covariant derivative can be used to describe the differential change of any local vector along a given path, because local vectors all transform in the same way. There are a number of useful forms equivalent to eqn (6.2). Firstly s can be replaced by $c\tau$ where τ is the proper time measured on a clock travelling the same path. When the path is that of light some alternative parameter to τ can always be defined to specify the path length consistently. Thus the covariant derivative can be rewritten in terms of a general path parameter λ which is linearly related to τ when the path is non light-like:

$$\frac{\mathrm{D}q^\mu}{\mathrm{D}\lambda} = \frac{\mathrm{d}q^\mu}{\mathrm{d}\lambda} + \Gamma^\mu_{\nu\rho}\frac{\mathrm{d}x^\nu}{\mathrm{d}\lambda}q^\rho. \tag{6.7}$$

For covector components

$$\frac{\mathrm{D}q_\mu}{\mathrm{D}\lambda} = \frac{\mathrm{d}q_\mu}{\mathrm{d}\lambda} - \Gamma^\nu_{\mu\rho}\frac{\mathrm{d}x^\rho}{\mathrm{d}\lambda}q_\nu.$$

Multiplying eqn (6.2) by $\partial s/\partial x^\nu$ gives another form:

$$\frac{\mathrm{D}q^\mu}{\mathrm{D}x^\nu} = \frac{\partial q^\mu}{\partial x^\nu} + \Gamma^\mu_{\nu\rho}q^\rho. \tag{6.8}$$

This can be written more compactly by using the subscript comma notation

$$q^\mu_{,\nu} = \frac{\partial q^\mu}{\partial x^\nu},$$

and introducing the subscript semicolon notation for the covariant derivatives:

$$q^\mu_{;\nu} = \frac{\mathrm{D}q^\mu}{\mathrm{D}x^\nu}.$$

Then eqn (6.8) becomes

$$q^\mu_{;\nu} = q^\mu_{,\nu} + \Gamma^\mu_{\nu\rho}q^\rho.$$

When a vector is *parallel transported* along a path it remains the same at each point on its path; thus

$$\frac{\mathrm{D}q^{\mu}}{\mathrm{D}s} = 0, \tag{6.9}$$

i.e.

$$\frac{\mathrm{d}q^{\mu}}{\mathrm{d}s} = -\Gamma^{\mu}_{\nu\rho}q^{\rho}\frac{\mathrm{d}x^{\nu}}{\mathrm{d}s},$$

i.e.

$$\mathrm{d}q^{\mu} = -\Gamma^{\mu}_{\nu\rho}q^{\rho}\,\mathrm{d}x^{\nu} \tag{6.10}$$

is consistent with eqn (6.1).

6.4. The principle of generalized covariance

This principle postulated by Einstein has two components. Firstly, physical laws must be expressible as tensor equations so that they remain valid under transformations to any accelerated frame. Secondly, when specialized to a frame in free fall the physical laws should reproduce the established laws consistent with SR. Fundamental to any study of dynamics is Newton's second law. It was expressed in a form consistent with SR in eqn (1.2), i.e.

$$F^{\mu} = \frac{\mathrm{d}p^{\mu}}{\mathrm{d}\tau}$$

where F^{μ} is the four vector force. Although the right hand side is not a valid tensor, we have just learnt how to write a covariant derivative that is a valid tensor. Replacing the derivative on the right hand side of eqn (1.2) by a covariant derivative yields

$$F^{\mu} = \frac{\mathrm{D}p^{\mu}}{\mathrm{d}\tau}. \tag{6.11}$$

This is a valid tensor equation *and* reduces to eqn (1.2) in a frame in free fall, so that it satisfies the principle of generalized covariance. Any other physical law consistent with SR in a frame in free fall can be converted into a physical law valid in accelerating frames by replacing space–time derivatives by the equivalent covariant derivatives. Clearly we have here a very powerful tool.

Using subscript comma and subscript semicolon notation a standard space–time derivative is written $p^{\mu}{}_{,\nu}$ whilst a covariant derivative is written $p^{\mu}{}_{;\nu}$. Hence the procedure for converting an equation so that it becomes valid in accelerating frames can be expressed pithily as: 'replace commas by semicolons'.

6.5. The geodesic equation

The simplest dynamical question that can be posed for curved space–time is to ask what the motion of a test body is in free fall, i.e. under gravitational forces alone. The gravitational forces are the manifestation of space–time curvature due to the presence of matter. Therefore in eqn (6.11) F^μ must be set to zero:

$$\frac{\mathrm{D}p^\mu}{\mathrm{D}\tau} = 0. \tag{6.12}$$

Referring to Section 6.3 we see that p^μ is being parallel transported. In addition p^μ, being the test body's momentum, points along the path. p^μ is therefore being parallel transported along itself, which is the prescription given in Section 3.3 for generating a geodesic! Hence the geometric view of the path of a body in free fall is that it follows a geodesic in space–time. Equation (6.12) is known as the *geodesic equation.* It can be converted to a useful equivalent form by making the replacement

$$p^\mu = m\frac{\mathrm{d}x^\mu}{\mathrm{d}\tau}$$

where m is the rest mass of the test body. Then the geodesic equation becomes

$$\frac{\mathrm{d}^2x^\mu}{\mathrm{d}\tau^2} + \Gamma^\mu_{\nu\rho}\frac{\mathrm{d}x^\nu}{\mathrm{d}\tau}\frac{\mathrm{d}x^\rho}{\mathrm{d}\tau} = 0 \tag{6.13}$$

If we wish to extend the equation to describe geodesics followed by light, we write

$$\mathrm{d}^2x^\mu/\mathrm{d}\lambda^2 + \Gamma^\mu_{\nu\rho}(\mathrm{d}x^\nu/\mathrm{d}\lambda)(\mathrm{d}x^\rho/\mathrm{d}\lambda) = 0 \tag{6.14}$$

where λ is the path parameter discussed in Section 6.3. Notice that the geodesic eqn (6.13) does not depend on the mass of the test body. This shows that the weak equivalence principle is already built into the very structure of Riemann space–time.

In Section 3.3 an alternative definition given for a geodesic was of a curve across space with no component of curvature in that space. This property of a geodesic is inherent in the geodesic equation, as we now show. The differential $\mathrm{d}x^\mu/\mathrm{d}s$ is a valid vector so we can write

$$\frac{\mathrm{D}x^\mu}{\mathrm{D}s} = \frac{\mathrm{d}x^\mu}{\mathrm{d}s}$$

Then the geodesic equation becomes

$$\frac{\mathrm{D}^2x^\mu}{\mathrm{D}s^2} = 0. \tag{6.15}$$

In geometrical terms the $\mathrm{D}x^\mu/\mathrm{D}s$ are the gradient components and the $\mathrm{D}^2x^\mu/\mathrm{D}s^2$ the curvature components of the path. Therefore the geodesic equation implies that a geodesic has no component of curvature in space–time.

6.6. Transformation to a frame in free fall

It has been argued that the space–time we inhabit is a Riemann space, and that locally space–time in free fall is the space–time of special relativity. In this section we check that starting at any event y in space–time it is always possible to transform to a frame for which locally

$$\left.\begin{aligned} g_{\mu\nu} &= \eta_{\mu\nu} \\ \text{and} \qquad g_{\mu\nu,\rho} &= 0 \end{aligned}\right\} \text{ at } y \tag{6.16}$$

where $\eta_{\mu\nu}$ is the Minkowski metric of eqn (4.1) or (4.2). Taking the Gaussian coordinates to be x'^{μ} at y we assert that the coordinates x^{μ} for which eqn (6.16) holds are given by a transformation

$$x'^{\mu} = x^{\mu} + G^{\mu}{}_{\nu\tau}\frac{x^{\nu}x^{\tau}}{2},$$

where the $G^{\mu}{}_{\nu\tau}$ are some constants. Differentiating this expression with respect to x^{α} and then again with respect to x^{β} gives

$$\frac{\partial x'^{\mu}}{\partial x^{\alpha}} = \delta^{\mu}{}_{\alpha} + G^{\mu}{}_{\alpha\tau}x^{\tau}$$

and

$$\frac{\partial^2 x'^{\mu}}{\partial x^{\alpha}\partial x^{\beta}} = G^{\mu}{}_{\alpha\beta}.$$

We can choose y to be the origin of the x^{μ} coordinates so that

$$\frac{\partial x'^{\mu}}{\partial x^{\alpha}} = \delta^{\mu}{}_{\alpha}.$$

Now

$$g_{\mu\nu} = \left(\frac{\partial x'^{\alpha}}{\partial x^{\mu}}\right)\left(\frac{\partial x'^{\beta}}{\partial x^{\nu}}\right)g'_{\alpha\beta}$$

which when differentiated with respect to x^{σ} gives

$$\begin{aligned} g_{\mu\nu,\sigma} = {} & (\partial^2 x'^{\alpha}/\partial x^{\mu}\partial x^{\sigma})(\partial x'^{\beta}/\partial x^{\nu})g'_{\alpha\beta} \\ & + (\partial x'^{\alpha}/\partial x^{\mu})(\partial^2 x'^{\beta}/\partial x^{\nu}\partial x^{\sigma})g'_{\alpha\beta} \\ & + (\partial x'^{\alpha}/\partial x^{\mu})(\partial x'^{\beta}/\partial x^{\nu})g'_{\alpha\beta,\sigma}. \end{aligned}$$

When the values of the differentials at y are substituted in this expression we have at y

$$g_{\mu\nu,\sigma} = G^{\alpha}{}_{\mu\sigma}g'_{\alpha\nu} + G^{\beta}{}_{\nu\sigma}g'_{\mu\beta} + g'_{\mu\nu,\sigma}.$$

If locally at y the second part of eqn (6.16) is to be satisfied after transforming to the unprimed coordinates we must therefore have

$$g'_{\mu\nu,\sigma} + G^{\sigma}{}_{\mu\sigma} g'_{\alpha\nu} + G^{\beta}{}_{\nu\sigma} g'_{\mu\beta} = 0.$$

Referring to Section 6.2 it is clear that the required transformation can be achieved if we set $G^{\alpha}{}_{\mu\sigma}$ equal to the value of $-\Gamma^{\alpha}{}_{\mu\sigma}$ at y. The unprimed $g_{\mu\nu}$ will not necessarily be diagonal as it should be if it is to satisfy the first part of eqn (6.16). However some linear transformation to new coordinates

$$\bar{x}^{\mu} = a^{\mu}{}_{\nu} x^{\nu}$$

with constant coefficients $a^{\mu}{}_{\nu}$ will achieve this result at y; without affecting $g_{\mu\nu,\sigma}$. Thus we have verified that a transformation to a frame in free fall is always possible.

6.7. Geodesics as stationary paths

From the geometric view geodesics are the straightest paths between space–time events; physically they are the paths of test bodies in free fall. A third equivalent property of geodesics is that they are 'stationary' paths between space–time events. This final property is useful because it simplifies the calculation of general relativistic orbits. In flat space a straight line is the shortest path between two points: $S_{\mathrm{AC}} < S_{\mathrm{AB}} + S_{\mathrm{AC}}$ as in Fig. 6.3(a). The corresponding case for Minkowski space is shown in Fig. 6.3(b) for AC being time-like. A suitable Lorentz transformation makes the time axis lie along AC and then

$$S_{\mathrm{AC}} = ct_{\mathrm{AC}}$$

$$S_{\mathrm{AB}} = (c^2 t_{\mathrm{AB}}^2 - x^2)^{1/2}$$

$$S_{\mathrm{BC}} = (c^2 t_{\mathrm{BC}}^2 - x^2)^{1/2},$$

where x is the spatial coordinate of B. Then

$$S_{\mathrm{AC}} > S_{\mathrm{AB}} + S_{\mathrm{BC}}.$$

The straight-line path is therefore the longest path in Minkowski space. A description that covers both cases is to say that geodesics are stationary paths; this means that any small deviation of path from a geodesic produces no change in the length to first order in the deviation. Expressed formally this requirement is

$$\delta S = 0 \tag{6.17a}$$

where the integrated path length is

$$S = \int (g_{\mu\nu}\, \mathrm{d}x^{\mu}\, \mathrm{d}x^{\nu})^{1/2}. \tag{6.17b}$$

If the path length is parametrized by the proper path length ds then

$$S = \int \left(g_{\mu\nu} \frac{\mathrm{d}x^{\mu}}{\mathrm{d}s} \frac{\mathrm{d}x^{\nu}}{\mathrm{d}s} \right)^{1/2} \mathrm{d}s.$$

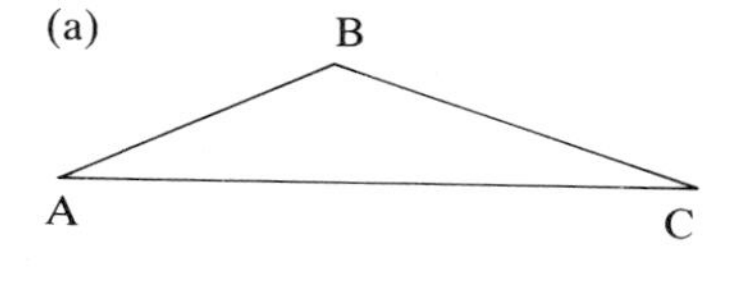

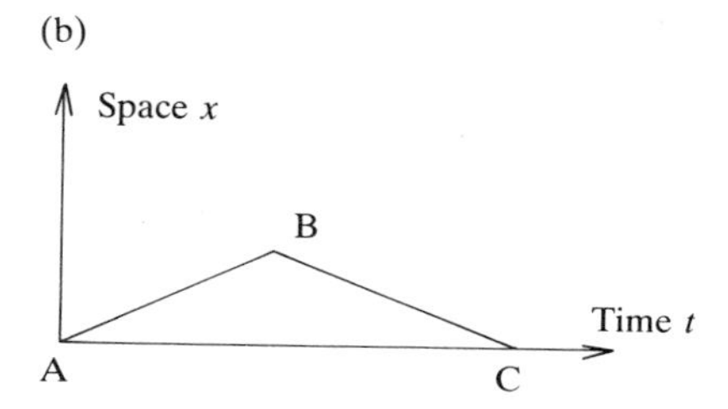

Fig. 6.3 (a) A straight line path AC in space and another longer path ABC; (b) a straight line path in space–time AC, and a displaced path ABC.

This integrand is unity all along the path so that if s is stationary then so too is a simpler integral

$$\delta I = 0 \tag{6.18a}$$

where

$$I = \int \left(g_{\mu\nu} \frac{\mathrm{d}x^\mu}{\mathrm{d}s} \frac{\mathrm{d}x^\nu}{\mathrm{d}s} \right) \mathrm{d}s. \tag{6.18b}$$

Equations (6.18a) and (6.18b) set a variational problem that is solved in the standard way in Appendix A. It is shown in Appendix A that eqn (6.18) is the integral form of the geodesic equation, and is entirely equivalent to the differential form of eqn (6.8).

It is as well to appreciate that a body in free fall from a fixed starting event will follow different geodesics if it is given different starting velocities. These geodesics will lie inside the forward light cone through the starting event; they are *time-like* with $\int \mathrm{d}s^2 > 0$. When the test body is a photon the path integral $\int \mathrm{d}s^2 = 0$ which defines a *null* geodesic. Finally the *space-like* geodesics that have $\int \mathrm{d}s^2 < 0$ would correspond to motion with velocity greater than c. Neither material particles nor light can follow space-like geodesics; however, these geodesics are useful in setting up coordinate frames that span space–time.

6.8. Familiar examples

In this section some familiar examples of gravitational and inertial forces will be used to bring out the correspondence between these forces and the metric connections. First let us establish how the geodesic equation simplifies when the motion can be described approximately by classical mechanics. Then the velocities must be small compared with c, the gravitational fields will be small and their rates of change will be slow. The geodesic equation is

$$\frac{d^2x^\mu}{ds^2} + \Gamma^\mu_{\alpha\beta}\frac{dx^\alpha}{ds}\frac{dx^\beta}{ds} = 0.$$

If the velocities are small, then

$$dx^0 = c\,dt \gg dx^i.$$

Therefore dx^0/ds is much larger than the derivatives of space components and in addition $dx^0/ds \approx 1$. Thus the dominant components of the geodesic equation are

$$d^2x^\mu/dt^2 + c^2\Gamma^\mu_{00} = 0,$$

provided that the Γ^μ_{00} are not all negligible. The time component

$$\Gamma^0_{00} = g_{00}\frac{(dg_{00}/dt)}{2c}$$

is negligible in the classical limit of fields which vary slowly with time. Thus, provided that the Γ^i_{00} are not negligible the dominant components of the geodesic equation in the classical limit are

$$\frac{d^2x^i}{dt^2} + c^2\Gamma^i_{00} = 0 \qquad (6.19)$$

Now if there is a constant acceleration g in the i direction we would write classically

$$\frac{d^2x^i}{dt^2} = g.$$

A comparison of the last two equations yields

$$\Gamma^i_{00} = -g/c^2 \qquad (6.20)$$

and shows the equivalence between an inertial force of classical mechanics and the metric connections. Both can be made to vanish by an appropriate coordinate transformation. If g is taken to be the local gravitational acceleration then we can see that the metric connections correspond to the components of the gravitational acceleration. This is a reasonable result when it is recalled that the metric components correspond to the classical gravitational potential (Section 4.3). More generally, if there is a gravitational potential φ, we have classically

$$\frac{d^2x^i}{dt^2} = -\frac{\partial\varphi}{\partial x^i}.$$

Comparison with eqn (6.19) shows that

$$\Gamma^i_{00} = (\partial\varphi/\partial x^i)/c^2.$$

When the linear acceleration is unimportant or zero then the Γ^i_{00} are negligible. In this case components of the geodesic equation other than those given by eqn (6.19) become the most important. So we come to another familiar example: the inertial acceleration observed in a frame rotating at constant angular velocity $\boldsymbol{\omega}$. It is well known that the absolute velocity is given by

$$v = \frac{\mathrm{d}\boldsymbol{r}}{\mathrm{d}t} + \boldsymbol{\omega} \times \boldsymbol{r}$$

where $\mathrm{d}\boldsymbol{r}/\mathrm{d}t$ is the velocity measured relative to the rotating frame at a vector distance $\boldsymbol{r}$ from the axis of rotation. This equation can be interpreted as an expression for the covariant derivative

$$v^\mu = \mathrm{D}x^\mu/\mathrm{D}\tau = \mathrm{d}x^\mu/\mathrm{d}\tau + \Gamma^\mu_{\alpha 0} x^\alpha.$$

Of course in the Newtonian limit of low velocities t can be replaced by τ. Making a term by term comparison of these last two equations we have

$$-\Gamma^1_{20} = \Gamma^2_{10} = \omega_3,$$
$$-\Gamma^3_{10} = \Gamma^1_{30} = \omega_2,$$
$$-\Gamma^2_{30} = \Gamma^3_{20} = \omega_1.$$

All the other metric connections vanish. The geodesic equation for the rotating frame

$$\frac{\mathrm{D}v^\mu}{\mathrm{D}t} = 0$$

therefore reduces to

$$\frac{\mathrm{d}v^\mu}{\mathrm{d}t} + \Gamma^\mu_{\alpha 0} v^\alpha = 0.$$

Substituting for v^μ (the absolute velocity) gives

$$\frac{\mathrm{d}^2 x^\mu}{\mathrm{d}t^2} + \Gamma^\mu_{\alpha 0}\frac{\mathrm{d}x^\alpha}{\mathrm{d}t} + \Gamma^\mu_{\alpha 0}\left(\frac{\mathrm{d}x^\alpha}{\mathrm{d}t} + \Gamma^\alpha_{\rho 0} x^\rho\right) = 0.$$

Therefore

$$\frac{\mathrm{d}^2 x^\mu}{\mathrm{d}t^2} = -2\Gamma^\mu_{\alpha 0}\frac{\mathrm{d}x^\alpha}{\mathrm{d}t} - \Gamma^\mu_{\alpha 0}\Gamma^\alpha_{\rho 0} x^\rho.$$

The 1-component of this equation is

$$\frac{\mathrm{d}^2 x^1}{\mathrm{d}t^2} = 2\omega_3\frac{\mathrm{d}x^2}{\mathrm{d}t} - 2\omega_2\frac{\mathrm{d}x^3}{\mathrm{d}t} - \omega_3\omega_1 x^3 + \omega_3^2 x^1 - \omega_2\omega_1 x^2 + {\omega_2}^2 x^1.$$

Collecting this and the other spatial terms (2, 3) into one equation gives

$$\frac{\mathrm{d}^2\boldsymbol{r}}{\mathrm{d}t^2} = -2\left(\boldsymbol{\omega} \wedge \frac{\mathrm{d}\boldsymbol{r}}{\mathrm{d}t}\right) - \boldsymbol{\omega} \wedge (\boldsymbol{\omega} \wedge \boldsymbol{r}).$$

The terms on the right-hand side are just the standard expressions for the Coriolis acceleration and centrifugal acceleration in a rotating frame. This again shows how the metric connections correspond to the inertial accelerations and forces of classical mechanics. This example illustrates another useful point, i.e. that a frame in free fall as generally understood must be a frame free of any rotation about the geodesic path; clearly metric connections can only vanish if both linear acceleration and acceleration due to rotation vanish. This is not quite the end of the matter. It is always possible, even in free fall, to obtain non-vanishing metric connections by choosing Gaussian rather than Cartesian coordinates. A simple example which illustrates this point is to choose polar coordinates (r, θ) to describe motion in a plane. The metric equation for free fall is

$$ds^2 = c^2\,dt^2 - dr^2 - r^2\,d\theta^2$$

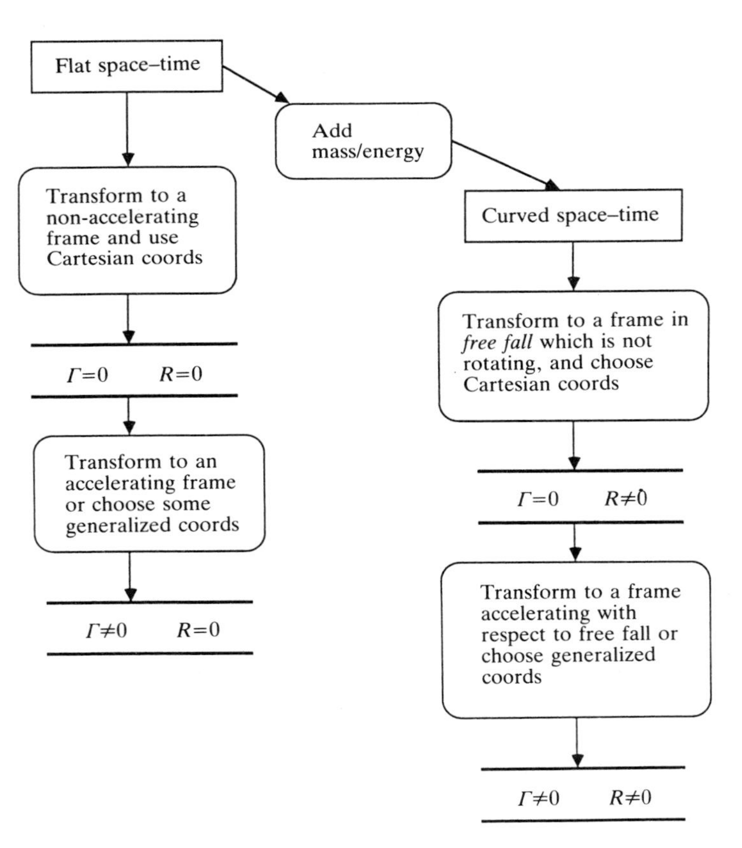

Fig. 6.4. Summary of the effects of both space–time curvature and coordinate transformations on the metric connection and the Riemann curvature tensor.

so that $g_{00} = g^{00} = 1$, $g_{11} = g^{11} = -1$ and $g_{22} = 1/g^{22} = -r^2$. One derivative does not vanish:

$$g_{22,1} = -2r$$

whence $\Gamma^2_{12} = \Gamma^2_{21} = 1/r$ and $\Gamma^1_{22} = -r$. Substituting these connections into the radial and θ components of the geodesic eqn (6.13) yields:

$$\frac{d^2r}{d\tau^2} - r\left(\frac{d\theta}{d\tau}\right)^2 = 0$$

$$\frac{d^2\theta}{d\tau^2} + \frac{2}{r}\frac{d\theta}{d\tau}\frac{dr}{d\tau} = 0.$$

In the Newtonian limit of small velocities $\tau \approx t$ and so these equations become

$$\ddot{r} - r\dot{\theta}^2 = 0 \qquad \text{and} \qquad r\ddot{\theta} + 2\dot{r}\dot{\theta} = 0.$$

These geodesic equations are quite familiar. On the left-hand sides are the well-known expressions for the radial and tangential components of acceleration. As expected both are zero for motion in the absence of any external force.

It is emphasized here that there is no rotation in a frame in free fall. Throughout what follows either Cartesian or spherical polar coordinates are used when analysing motion in free fall; the latter choice is made whenever the system studied has spherical symmetry. Figure 6.4 is based on the comments of the last paragraphs.

7
Einstein's theory II

The general technique for converting an equation valid for SR into one valid in all frames, namely replacing normal derivatives by covariant derivatives, cannot be applied to gravitation. It cannot be applied because we lack the starting point—a law of gravitation consistent with SR. Newton's law of gravitation has no explicit time dependence, implying that gravitational effects are transmitted instantaneously to all parts of the Universe; however, it is a basic postulate of SR that signals travel no faster than the speed of light. There is an underlying difficulty of principle in asking for a law of gravitation consistent with SR: SR applies to flat space–time while gravitational forces are a manifestation of its curvature. Einstein perceived that there must be a direct link between the distribution of mass/energy and the curvature of space–time and that this link must be expressible in tensor form. Einstein also recognized that the stress-energy tensor provided the appropriate tensor description for the distribution and flow of energy in space–time. Later he identified a curvature tensor (the Einstein tensor) having formal properties which match those of the stress-energy tensor. Einstein simply equated these two tensors, making 'curvature' at a given point in space–time proportional to 'energy density' at the same point. This is the essence of Einstein's equation.

In Section 7.1 the Riemann curvature tensor is introduced; this is a tensor which provides a complete description of curvature in multi-dimensional spaces. Its relationship both to the Gaussian curvature of two-dimensional surfaces and to tidal forces is discussed. In Section 7.2 attention is focused on the stress-energy tensor $T_{\mu\nu}$ and its properties. The conjecture that the stress-energy tensor is proportional to some curvature tensor leads to the selection for this role of a unique contraction of the Riemann tensor (Section 7.3) called the Einstein tensor $G_{\mu\nu}$. Then the Einstein equation is

$$G_{\mu\nu} = \frac{8\pi G T_{\mu\nu}}{c^4}. \tag{7.1}$$

The choice of the constant of proportionality $8\pi G/c^4$ will be justified in Section 7.4 by taking the classical limit and showing that Newton's law of gravitation is then obtained. In Appendix D Einstein's equation is applied to the region outside a spherically symmetric mass distribution, yielding the Schwarzschild metric.

7.1. The Riemann curvature tensor

The curvature of a two-dimensional surface is quantified by giving the value of the Gaussian curvature at each point. However, in higher-dimensional spaces each geodesic surface has its own Gaussian curvature and the description of curvature in terms of Gaussian curvatures becomes unwieldy. Riemann developed more elegant techniques to analyse curvature in these higher-dimensional spaces. A full description of curvature at a given point in such a space is provided by a rank 4 tensor called the Riemann curvature tensor. It is not difficult to infer where the information on curvature must come from. First we note that by going to a frame in free fall at an event (point in space–time) x the space–time is made flat locally, and we have (eqn 6.16)

$$\left.\begin{aligned} g_{\mu\nu} &= \eta_{\mu\nu} \\ g_{\mu\nu,\rho} &= 0 \end{aligned}\right\} \text{ at } x.$$

In these circumstances neither the metric coefficients nor their first derivatives have anything to say about curvature. What distinguishes curved space–time from flat space–time is this: the frame in free fall at x differs from the frame in free fall at an event a small distance away, at $x + \Delta x$. The metric tensor and its derivatives at $x + \Delta x$ can be obtained by Taylor expansions around $g_{\mu\nu}(x)$ and $g_{\mu\nu,\rho}(x)$:

$$g_{\mu\nu}(x + \Delta x) = \eta_{\mu\nu} + \frac{1}{2} g_{\mu\nu,\rho\sigma} \Delta x^\rho \Delta x^\sigma$$

and

$$g_{\mu\nu,\rho}(x + \Delta x) = g_{\mu\nu,\rho\sigma} \Delta x^\sigma.$$

The change in $g_{\mu\nu}$ depends only on the *second* derivatives $g_{\mu\nu,\rho\sigma}$ at x, and so these derivatives must embody the curvature information.

In Section 4.3 the correspondence between the metric components and the Newtonian gravitational potential was noted, and later in Section 6.8 it was seen that the metric connection and hence the first derivatives of the metric components correspond to the gravitational acceleration. A tidal acceleration is the difference between gravitational acceleration at different locations so that the corresponding parameters in curved space–time are the *second* derivatives of the metric connections. There is therefore a fundamental link between the curvature of space–time and the existence of classical tidal forces.

The curvature tensor can be identified by using an approach that was introduced in Chapter 3. A vector is parallel transported around an infinitesimal closed path and the change that this operation produces in the vector is calculated. Figure 7.1 shows the route. AB is a vector path element da and BC is a vector path element db; the return path consists of the vector elements $-\mathrm{d}a$ (CD) and $-\mathrm{d}b$ (DA). Parallel transporting a vector v^α around the loop

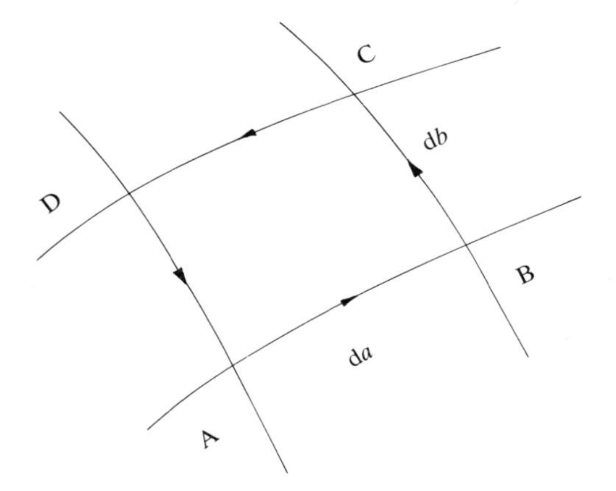

Fig. 7.1 A closed path in curved space–time around which a vector is parallel transported.

ABCDA induces changes which are calculated using eqn (6.10). The changes induced over each of the four segments of path are respectively

$$-\Gamma^{\alpha}_{\beta\nu}(x)v^{\nu}(x)\,\mathrm{d}a^{\beta}$$

$$-\Gamma^{\alpha}_{\beta\nu}(x+\mathrm{d}a)v^{\nu}(x+\mathrm{d}a)\,\mathrm{d}b^{\beta}$$

$$+\Gamma^{\alpha}_{\beta\nu}(x+\mathrm{d}b)v^{\nu}(x+\mathrm{d}b)\,\mathrm{d}a^{\beta}$$

and

$$+\Gamma^{\alpha}_{\beta\nu}(x)v^{\nu}(x)\,\mathrm{d}b^{\beta}.$$

$\Gamma^{\alpha}{}_{\beta\nu}$ and v^{β} are evaluated at the location indicated within parentheses. Collecting these terms yields the overall change suffered by v^{α}:

$$\mathrm{d}v^{\alpha} = \frac{\partial(\Gamma^{\alpha}_{\beta\nu}v^{\nu})}{\partial x^{\gamma}}\,\mathrm{d}b^{\gamma}\,\mathrm{d}a^{\beta} - \frac{\partial(\Gamma^{\alpha}_{\beta\nu}v^{\nu})}{\partial x^{\delta}}\,\mathrm{d}a^{\delta}\,\mathrm{d}b^{\beta}.$$

Using eqn (6.10) again we obtain

$$\begin{aligned}\mathrm{d}v^{\alpha} &= \mathrm{d}a^{\beta}\,\mathrm{d}b^{\gamma}(\Gamma^{\alpha}_{\beta\nu,\gamma}v^{\nu} - \Gamma^{\alpha}_{\beta\nu}\Gamma^{\nu}_{\sigma\gamma}v^{\sigma}) - \mathrm{d}a^{\delta}\,\mathrm{d}b^{\beta}(\Gamma^{\alpha}_{\beta\nu,\delta}v^{\nu} - \Gamma^{\alpha}_{\beta\nu}\Gamma^{\nu}_{\sigma\delta}v^{\sigma})\\ &= \mathrm{d}a^{\delta}\,\mathrm{d}b^{\gamma}\,v^{\beta}(\Gamma^{\alpha}_{\beta\delta,\gamma} - \Gamma^{\alpha}_{\beta\gamma,\delta} + \Gamma^{\alpha}_{\sigma\gamma}\Gamma^{\sigma}_{\beta\delta} - \Gamma^{\alpha}_{\sigma\delta}\Gamma^{\sigma}_{\beta\gamma}).\end{aligned}$$

In writing the last line use has been made of identities such as

$$\mathrm{d}a^{\beta}\Gamma^{\alpha}_{\beta\nu,\gamma} = \mathrm{d}a^{\delta}\Gamma^{\alpha}_{\delta\nu,\gamma}$$

which result from changing a repeated suffix, in this case $\beta \to \delta$. The expression for $\mathrm{d}v^{\alpha}$ can be written compactly as

$$\mathrm{d}v^{\alpha} = \mathrm{d}a^{\delta}\,\mathrm{d}b^{\gamma}\,v^{\beta}R^{\alpha}{}_{\beta\gamma\delta} \tag{7.2}$$

where $R^{\alpha}{}_{\beta\gamma\delta}$ is defined by

$$R^{\alpha}_{\beta\gamma\delta} = \Gamma^{\alpha}_{\beta\delta,\gamma} - \Gamma^{\alpha}_{\beta\gamma,\delta} + \Gamma^{\alpha}_{\sigma\gamma}\Gamma^{\sigma}_{\beta\delta} - \Gamma^{\alpha}_{\sigma\delta}\Gamma^{\sigma}_{\beta\gamma}. \tag{7.3}$$

All the vectors v, $\mathrm{d}a$, and $\mathrm{d}b$ appearing in (7.2) are quite arbitrary; $\mathrm{d}v$ is the difference between local vectors at the same place and is also a vector. Consequently we can use the quotient theorem to prove that $R^{\alpha}{}_{\beta\gamma\delta}$ is a tensor. This, the *Riemann curvature tensor*, quantifies space–time curvature. Specializing to a frame in free fall causes all the connections to vanish but not their derivatives, and eqn (7.3) collapses to

$$R^{\alpha}_{\beta\gamma\delta} = \Gamma^{\alpha}_{\beta\delta,\gamma} - \Gamma^{\alpha}_{\beta\gamma,\delta}. \tag{7.4}$$

Other associated forms of the Riemann tensor are also of interest, e.g.

$$R_{\alpha\beta\gamma\delta} = g_{\alpha\mu} R^{\mu}_{\beta\gamma\delta}.$$

Appendix B contains more details on the mathematical properties of the Riemann tensor. One useful result established there is that eqn (7.4) can be written as

$$2R_{\alpha\beta\gamma\delta} = g_{\alpha\delta,\beta\gamma} - g_{\beta\delta,\alpha\gamma} + g_{\beta\gamma,\alpha\delta} - g_{\alpha\gamma,\beta\delta} \tag{7.5}$$

in which the anticipated dependence of curvature on the second derivatives of the metric coefficients is made explicit. If space–time is flat everywhere these derivatives, and the Riemann tensor, will all vanish.

The behaviour of the Riemann tensor is contrasted with that of the metric connections in Fig. 6.4. A suitable choice of frame can always be made such that the connections and the acceleration vanish locally. This is not the case for the Riemann tensor: it vanishes wherever space–time is flat, whatever generalized coordinates are chosen. The Riemann tensor for space–time has 4^4 components; in Appendix B it is shown that thanks to the symmetry of this tensor only 20 components are linearly independent.

Another property of curved spaces is that two nearby and initially parallel geodesics do not continue parallel indefinitely, but converge or diverge depending on the local curvature. For example, a pair of lines of longitude are parallel at the equator but converge toward the poles. Suppose that a vector ξ is drawn in curved space–time linking points on two nearby geodesics at x and $x + \xi$. The two geodesics have equations

$$0 = \frac{\mathrm{d}^2 x^{\mu}}{\mathrm{d}\tau^2} + \Gamma^{\mu}_{\nu\lambda}(x)\frac{\mathrm{d}x^{\nu}}{\mathrm{d}\tau}\frac{\mathrm{d}x^{\lambda}}{\mathrm{d}\tau}$$

and

$$\begin{aligned} 0 &= \frac{\mathrm{d}^2(x+\xi)^{\mu}}{\mathrm{d}\tau^2} + \Gamma^{\mu}_{\nu\lambda}(x+\xi)\frac{\mathrm{d}(x+\xi)^{\nu}}{\mathrm{d}\tau}\frac{\mathrm{d}(x+\xi)^{\lambda}}{\mathrm{d}\tau} \\ &= \frac{\mathrm{d}^2 x^{\mu}}{\mathrm{d}\tau^2} + \frac{\mathrm{d}^2 \xi^{\mu}}{\mathrm{d}\tau^2} + \Gamma^{\mu}_{\nu\lambda}(x)\frac{\mathrm{d}x^{\nu}}{\mathrm{d}\tau}\frac{\mathrm{d}x^{\lambda}}{\mathrm{d}\tau} + \Gamma^{\mu}_{\nu\lambda,\rho}(x)\xi^{\rho}\frac{\mathrm{d}x^{\nu}}{\mathrm{d}\tau}\frac{\mathrm{d}x^{\lambda}}{\mathrm{d}\tau} + \Gamma^{\mu}_{\nu\lambda}(x)\frac{\mathrm{d}x^{\nu}}{\mathrm{d}\tau}\frac{\mathrm{d}\xi^{\lambda}}{\mathrm{d}\tau} \\ &\quad + \Gamma^{\mu}_{\nu\lambda}(x)\frac{\mathrm{d}\xi^{\nu}}{\mathrm{d}\tau}\frac{\mathrm{d}x^{\lambda}}{\mathrm{d}\tau} + \mathrm{O}(\xi^2) \text{ terms.} \end{aligned}$$

The difference between these equations, to first order in ξ, is

$$0 = \frac{d^2\xi^\mu}{d\tau^2} + \Gamma^\mu_{\nu\lambda,\rho}\xi^\rho \frac{dx^\nu}{d\tau}\frac{dx^\lambda}{d\tau} + \Gamma^\mu_{\nu\lambda}\left(\frac{dx^\nu}{d\tau}\right)\left(\frac{d\xi^\lambda}{d\tau}\right) + \Gamma^\mu_{\nu\lambda}\left(\frac{d\xi^\nu}{d\tau}\right)\left(\frac{dx^\lambda}{d\tau}\right)$$

where the connections and their differentials are all evaluated at x. Specializing to a frame in free fall at x, so that the connections vanish, we have

$$0 = \frac{d^2\xi^\mu}{d\tau^2} + \Gamma^\mu_{\nu\lambda,\rho}\xi^\rho \frac{dx^\nu}{d\tau}\frac{dx^\lambda}{d\tau}.$$

Also for the frame in free fall

$$\frac{D^2\xi^\mu}{D\tau^2} = \frac{d^2\xi^\mu}{d\tau^2} + \Gamma^\mu_{\rho\lambda,\nu}\xi^\rho \frac{dx^\nu}{d\tau}\frac{dx^\lambda}{d\tau}.$$

Eliminating $d^2\xi^\mu/d\tau^2$ from these last two expressions gives

$$0 = \frac{D^2\xi^\mu}{D\tau^2} - \Gamma^\mu_{\rho\lambda,\nu}\xi^\rho \frac{dx^\nu}{d\tau}\frac{dx^\lambda}{d\tau} + \Gamma^\mu_{\nu\lambda,\rho}\xi^\rho \frac{dx^\nu}{d\tau}\frac{dx^\lambda}{d\tau}.$$

Then using eqn (7.4) this becomes, after rearrangement,

$$0 = \frac{D^2\xi^\mu}{D\tau^2} + R^\mu_{\nu\rho\lambda}\xi^\rho \frac{dx^\nu}{d\tau}\frac{dx^\lambda}{d\tau} \tag{7.6}$$

which, being expressed entirely in terms of tensors, will hold true under general transformations of coordinates. Equation (7.6) is known as the *equation of geodesic deviation.*

We shall now briefly bring out the relationship between the Gaussian curvature and the Riemann curvature tensor. First a geodesic surface is constructed in space–time according to the prescription given at the end of Section 3.4. The geodesic surface is spanned by geodesics tangential to the 1- and 2-axes at a given point, and for simplicity the frame at this point is taken to be one which is in free fall. Then the equation for the geodesic deviation in the 1-direction resulting from a displacement in the 2-direction is

$$\frac{D^2\xi^1}{D\tau^2} = -R^1{}_{212}\xi^1\left(\frac{dx^2}{d\tau}\right)^2.$$

Recalling that $ds = c\,d\tau$ we can rewrite this last line as

$$\frac{D^2\xi^1}{Ds^2} = -R^1{}_{212}\xi^1\left(\frac{dx^2}{ds}\right)^2.$$

In free fall the coordinates can be chosen to make the metric diagonal, in which case $R^1{}_{212} = g^{1\alpha}R_{\alpha 212}$ reduces to

$$R^1{}_{212} = g^{11}R_{1212} = R_{1212}/g_{11}$$

and

$$\left(\frac{dx^2}{ds}\right)^2 = \frac{1}{g_{22}}.$$

Therefore the previous equation becomes

$$\frac{D^2\xi^1}{Ds^2} = -\left(\frac{R_{1212}}{g_{11}g_{22}}\right)\xi^1.$$

Again with a diagonal metric, the metric determinant

$$g = g_{11}g_{22} - g_{12}^2 = g_{11}g_{22}.$$

Thus in free fall we have

$$\frac{D^2\xi^1}{Ds^2} = -\left(\frac{R_{1212}}{g}\right)\xi^1. \tag{7.7}$$

Equation 7.7 can be compared with the earlier expression for geodesic deviation:

$$\frac{d^2\eta}{ds^2} = +K\eta, \tag{4.6}$$

taking account of the sign change in the spatial components of the metric when passing from Euclidean to Minkowski space. Therefore the Gaussian curvature of the geodesic surface under discussion is

$$K = -R_{1212}/g. \tag{7.8}$$

This is the tensor form of Gauss's 'Excellent' Theorem applied to one geodesic surface. The experimental determination of components of the Riemann tensor involves measuring the gradient of the gravitational field. A device to do this is called a gravity gradiometer. Very sensitive devices based on technical developments stimulated by attempts to detect variations of m_g/m_i between different materials are now in prospect.

The next stage along our road toward Einstein's general theory will be to discover how to present the density of mass/energy in a way compatible with SR, i.e. as a tensor. Once this result is in place it will be possible to discuss how to relate space–time curvature mathematically to mass/energy.

7.2. The stress-energy tensor

In SR the energy and the linear momentum are two aspects of a single entity, the four-momentum. They are connected to the rest mass through the relation given in Section 1.3:

$$E^2 - c^2p^2 = m^2c^4.$$

This intimate connection of mass, energy and momentum suggests that a general law of gravitation will contain not just mass but energy and momentum as well. In order to proceed we need to convert Newton's law of gravitation to its differential form. The attractive force on unit mass a distance r from a body of mass M is just

$$F = GM/r^2.$$

Integrating the flux of F out through a sphere centred on the mass gives

$$\int \boldsymbol{F}\cdot \mathrm{d}\boldsymbol{S} = -4\pi r^2 \frac{GM}{r^2} = -4\pi GM.$$

Now Stokes's theorem relates an integral over a closed surface to another integral over the volume enclosed:

$$\int \boldsymbol{F}\cdot \mathrm{d}\boldsymbol{S} = \int \boldsymbol{\nabla}\cdot \boldsymbol{F}\,\mathrm{d}V.$$

Making this replacement gives:

$$\int \boldsymbol{\nabla}\cdot \boldsymbol{F}\,\mathrm{d}V = -4\pi GM,$$

i.e.

$$\int \boldsymbol{\nabla}\cdot \boldsymbol{F}\,\mathrm{d}V = -4\pi G \int \rho\,\mathrm{d}V,$$

where ρ is the density of the material inside the volume. This result is valid for any choice of volume so that

$$\boldsymbol{\nabla}\cdot \boldsymbol{F} = -4\pi G\rho.$$

In terms of the gravitational potential defined by $\boldsymbol{F} = -\boldsymbol{\nabla}\varphi$ this becomes

$$\nabla^2\varphi = 4\pi G\rho, \tag{7.9}$$

where ρ is the rest local density of matter and φ is the local gravitational potential (gravitational potential energy per unit mass). In order to see how ρ can be generalized in SR we consider a simple sample of matter, namely a dust cloud. A cloud of dust in its rest frame S has energy density

$$\rho_0 c^2 = m_0 n_0 c^2,$$

where m_0 is the rest mass of the average dust grain and n_0 is the number of dust grains per unit volume. Viewed in a frame S′ moving with velocity $v = \beta c$ with respect to the cloud, each grain becomes more massive, and the volume containing a fixed number of grains is Lorentz contracted along the direction of motion. Explicitly

$$m_0 \Rightarrow m = m_0\gamma$$

and

$$n_0 \Rightarrow n = n_0 \gamma,$$

where $\gamma = 1/(1 - \beta^2)^{1/2}$. Thus

$$\rho_0 \Rightarrow \rho = \rho_0 \gamma^2.$$

Obviously ρc^2 is neither a scalar nor the component of a four-vector: if it were the first it would remain constant under the Lorentz transformation to S′; if it were the second the change would be linear in γ. The behaviour of ρ is, however, exactly that of the time–time component of a second-rank tensor $T^{\mu\eta}$:

$$T^{\mu\nu} = \rho_0 v^\mu v^\nu, \tag{7.10}$$

where v^μ is the four-vector velocity of the cloud. In the frame S only the time–time component of this tensor is non-zero; it is $(T^{00})_0 = \rho_0 c^2$. Under the transformation from S to S′

$$(T^{00})_0 \Rightarrow T^{00} = \gamma^2 (T^{00})_0.$$

The tensor $T^{\mu\nu}$ is called the *stress-energy* tensor. A definition of its components that applies equally to dust clouds or more complex systems is the following: $T^{\mu\nu}$ is the flow of the μ component of the four-momentum along the ν direction. Some examples should help to make this clearer.

(a) T^{00} is the energy density.
(b) cT^{0i} is the energy flow per unit area parallel to the *i* direction. In the case of the dust cloud this would constitute heat flow.
(c) T^{ii} is the flow of momentum component *i* per unit area in the *i* direction, i.e. the pressure across the *i* plane.
(d) T^{ij} is the flow of the *i* component of momentum per unit area in the *j* direction. This is another way of describing a component of the viscous drag across the *j* plane.
(e) $T^{i0}c$ is the density of the *i* component of momentum.

It is apparent that in SR the stress-energy tensor embodies a compact description of energy and momentum density. In contrast, the underlying connection between the energy, momentum, pressure, and heat flow is only latent in the classical (Newtonian) view. Referring to the dust cloud it can be seen that if a Lorentz transformation is made to another frame, the spatial components of $T^{\mu\nu}$ generally receive contributions from $(T^{00})_0$: thus the spatial components must also affect the curvature of space–time. Consequently *all* the terms appearing in the stress-energy tensor must help to warp the fabric of space–time! Herein lies a paradox for the classical view. It is well known that the pressure inside a star resists the gravitational collapse, and yet it now looks as if pressure can, by virtue of contributing to a component of the stress-energy tensor, hasten gravitational collapse. This is indeed the case, and pressure contributes to the contraction of sufficiently massive stars to black holes (see Section 9.4).

Conservation laws for energy and momentum take particularly simple forms when expressed in terms of the stress-energy tensor. The conservation of energy offers a good example of the way that this happens. Figure 7.2 shows a cube of edge length l with its edges parallel to the axes Ox, Oy, and Oz in a medium whose stress-energy tensor is $T^{\mu\nu}$. The rate of change in the energy content of the box is

$$l^3 \frac{\partial T^{00}}{\partial t}.$$

This change is produced by the net energy inflow through the six faces of the cube. The energy flow through the faces at $x = x$, and $x = x + l$ is respectively

$$l^2 c T^{01}(x) \qquad \text{inward per unit time}$$

$$l^2 c T^{01}(x + l) \qquad \text{outward per unit time.}$$

The net flow inward from these two faces is

$$l^2 c[T^{01}(x) - T^{01}(x + l)] = -l^3 c \frac{\partial T^{01}}{\partial x}.$$

There are similar contributions from the other pairs of faces:

$$-l^3 c\left(\frac{\partial T^{02}}{\partial y}\right) \qquad \text{and} \qquad -l^3 c\left(\frac{\partial T^{03}}{\partial z}\right).$$

Summing the three contributions gives the total inflow, and so we have

$$l^3 \frac{\partial T^{00}}{\partial t} = -l^3 c\left(\frac{\partial T^{01}}{\partial x} + \frac{\partial T^{02}}{\partial y} + \frac{\partial T^{03}}{\partial z}\right),$$

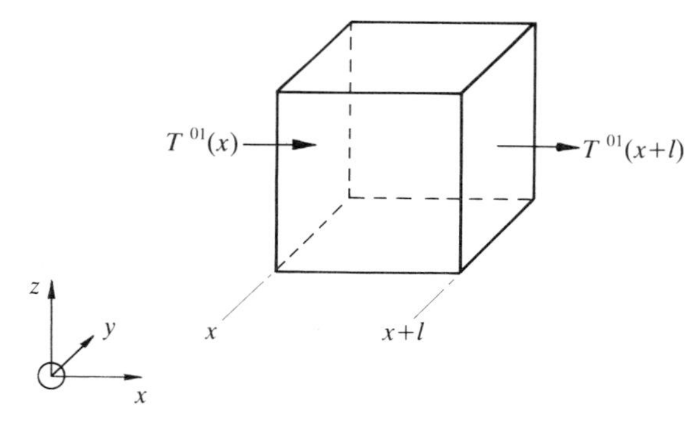

Fig. 7.2 A cube of side l located in a material whose stress-energy tensor is $T^{\alpha\beta}$.

which can be rearranged to become

$$\frac{\partial T^{00}}{\partial x^{0}} + \frac{\partial T^{0i}}{\partial x^{i}} = 0,$$

that is,

$$\frac{\partial T^{0\alpha}}{\partial x^{\alpha}} = 0.$$

This last result can be expressed in a more compact form using the subscript comma notation for space–time derivatives:

$$T^{0\alpha}{}_{,\alpha} = 0. \tag{7.11}$$

This type of derivative is known as the *divergence* of $T^{0\alpha}$; the derivative is taken with respect to the same space–time component as appears in the tensor (here α), and the sum over the expression on the left-hand side is made for all values of α. A parallel procedure applied to the conservation of linear momentum yields the three equations

$$T^{i\alpha}{}_{,\alpha} = 0 \qquad \text{for } i = 1, 2, \text{ and } 3. \tag{7.12}$$

Finally the conservation laws (7.11) and (7.12) can be combined into a single equation:

$$T^{\beta\alpha}{}_{,\alpha} = 0 \qquad \text{for } \beta = 0, 1, 2, \text{ and } 3.$$

This equation summarizes all the conservation laws of four-momentum in SR. Put into words they require that the divergences of the stress–energy tensor vanish everywhere. This result from SR can be converted to a form of general validity in curved space–time if the simple derivatives are replaced by covariant derivatives. Thus the laws of conservation of energy and momentum in curved space–time take the form

$$T^{\beta\alpha}{}_{;\alpha} = 0, \tag{7.13}$$

where the subscript semicolon indicates a covariant derivative. This can be written in more detail as

$$\frac{\partial T^{\beta\alpha}}{\partial x^{\alpha}} + \Gamma^{\alpha}{}_{\mu\alpha} T^{\beta\mu} + \Gamma^{\beta}{}_{\mu\beta} T^{\mu\alpha} = 0.$$

The last line illustrates that in the covariant derivative of a tensor of second order or higher there is one term for each index. The properties of the energy–momentum tensor are now summarized:

(a) it vanishes in the absence of matter;
(b) it is of second rank;
(c) its divergences all vanish;
(d) it is symmetric, i.e. $T^{\mu\nu} = T^{\nu\mu}$.

7.3. Einstein's equation

Einstein identified the stress–energy tensor as the *source* of space–time curvature and suggested the simplest possible relationship between it and the curvature:

$$KT_{\mu\nu} = G_{\mu\nu},$$

where $G_{\mu\nu}$ is a tensor describing space–time curvature and K is some scalar constant whose magnitude determines how effective the energy density is in distorting space–time. An immediate consequence of this ansatz is that $G_{\mu\nu}$ should be a symmetric divergenceless second-rank tensor to match the stress-energy tensor. The Riemann curvature tensor (rank 4) quantifies space–time curvature, and so it is reasonable to expect that $G_{\mu\nu}$ should be a contraction of the Riemann curvature tensor. A second-rank tensor can easily be constructed by contraction:

$$R_{\beta\delta} = R^{\alpha}{}_{\beta\alpha\delta} = g^{\alpha\sigma}R_{\sigma\beta\alpha\delta}. \tag{7.14}$$

which is called the Ricci tensor. Using Appendix B it can be shown that the contractions $R^{\alpha}{}_{\alpha\beta\delta}$ and $R^{\alpha}{}_{\beta\gamma\alpha}$ give equivalent results: the Ricci tensor is thus the *unique* contraction of the Riemann tensor. The Ricci tensor has a non-zero divergence which can be removed by a simple subtraction. The result is the divergenceless second-rank Einstein tensor

$$G_{\beta\delta} = R_{\beta\delta} - g_{\beta\delta}R/2 \tag{7.15}$$

where R is the Ricci scalar $R = g^{\beta\delta}R_{\beta\delta}$. The proof that the Einstein tensor is both symmetric and divergenceless is given in Appendix B. Like its progenitors, the Riemann and Ricci tensors, the Einstein tensor vanishes in the absence of any material to warp space–time. Referring to the summary of the properties of the stress-energy tensor given at the end of the previous section, we see that the Einstein tensor matches these precisely. With this tensor we can now rewrite Einstein's ansatz as

$$G_{\alpha\beta} = \frac{8\pi G}{c^4}T_{\alpha\beta}. \tag{7.16}$$

The value of the constant $8\pi G/c^4$ is fixed by the requirement that, in the limit of weak slowly varying gravitational fields, the Einstein equation should reduce to Newton's law of gravitation. This will be shown explicitly in Section 7.4. Note that the G appearing on the right-hand side of (7.16) is the usual gravitational constant and not some contraction of a tensor. As with all such tensor equations the equality will remain true if we simultaneously raise one or more indices on both sides of the equation.

When Einstein's equation was applied to calculate the behaviour of the Universe on the large scale it was apparent that it favoured an expanding

Universe; however, this conflicted with the prejudice, then current, that the Universe was static. Einstein found that he could extend his equation in only one way, by adding a term $\Lambda g_{\alpha\beta}$ so that

$$G_{\alpha\beta} - \Lambda g_{\alpha\beta} = \frac{8\pi G}{c^4} T_{\alpha\beta} \tag{7.17}$$

where Λ is a universal constant called the *cosmological constant*. As we can see, it would lead to curvature of space–time in the absence of all matter and radiation ($T_{\alpha\beta} = 0$). It is therefore possible by choosing Λ appropriately to obtain a static Universe. By 1931 Hubble and Humason had convincingly demonstrated that the Universe is currently expanding and Einstein was then happy to disown the cosmological constant. Curiously, the cosmological constant is once again in vogue. Quantum theory has taught us that the vacuum is not a featureless void but is continuously disturbed by particle–anti-particle pair creation and annihilation and that this endows the vacuum with energy. A non-zero cosmological constant no longer seems at all unreasonable.

An alternative form of Einstein's equation is also useful. Rewriting eqn (7.17) we obtain

$$R_{\alpha\beta} - \frac{R g_{\alpha\beta}}{2} = \frac{8\pi G T_{\alpha\beta}}{c^4} + \Lambda g_{\alpha\beta}. \tag{7.18}$$

First we raise one index and contract it with the remaining covector index to give

$$R - \frac{R g^{\beta}{}_{\beta}}{2} = \frac{8\pi G T^{\beta}{}_{\beta}}{c^4} + \Lambda g^{\beta}{}_{\beta}.$$

Now $g^{\beta}{}_{\beta}$ is a scalar and will have the same value in all frames including that in free fall; explicitly we have

$$g^{\beta}{}_{\beta} = g^{\beta\alpha} g_{\alpha\beta} = \eta^{\beta\alpha} \eta_{\alpha\beta} = 4.$$

Thus

$$-R = \frac{8\pi G T^{\mu}{}_{\mu}}{c^4} + 4\Lambda.$$

Substituting for R in eqn (7.18) gives

$$R_{\alpha\beta} = \frac{8\pi G (T_{\alpha\beta} - T^{\mu}{}_{\mu} g_{\alpha\beta}/2)}{c^4} - \Lambda g_{\alpha\beta}. \tag{7.19}$$

We can see that in empty space–time Einstein's equation simplifies to

$$R_{\alpha\beta} = -\Lambda g_{\alpha\beta},$$

and that $R_{\alpha\beta}$ vanishes there if the cosmological constant is zero. This result will be useful in later discussions of the passage of gravitational waves through otherwise empty space–time.

7.4 The Newtonian limit

It is instructive to check that in the limit of weak slowly-varying gravitational fields Einstein's equation reduces to Newton's law of gravitation. This will show how the new variables of Einstein's general theory are related to the more familiar classical variables. The starting point is Einstein's equation (7.19). In the classical limit when the gravitational field is weak and slowly varying

$$g_{\mu\nu} = \eta_{\mu\nu} + h_{\mu\nu}$$

where all the components of the tensor h are much less than unity. If we choose Cartesian coordinates $\eta_{00} = +1$, $\eta_{11} = \eta_{22} = \eta_{33} = -1$. In the same limit velocities are small ($\ll c$) and the spatial components of momenta are much less than energies. The dominant term in the stress-energy tensor is therefore the energy density T_{00}. Thus the important part of Einstein's equation in the classical limit is

$$R_{00} = \frac{8\pi G(T_{00} - T^{0}{}_{0}g_{00}/2)}{c^4} - \Lambda g_{00}. \tag{7.20}$$

In evaluating this expression we first note that the metric connections are linear in h; hence to a first approximation in h the form (7.4) or (7.5) can be used for the Riemann curvature tensor rather than (7.3). Then the corresponding approximation for the Ricci tensor is that given in eqn B.7:

$$R_{00} = \eta^{\alpha\sigma}\frac{h_{\sigma 0,0\alpha} - h_{00,\sigma\alpha} + h_{0\alpha,\sigma 0} - h_{\sigma\alpha,00}}{2}.$$

In the classical (slow-moving) limit it is possible to ignore the time compared with spatial derivatives so that the last equation reduces to

$$R_{00} = -\frac{\eta^{ij}h_{00,ij}}{2} = \frac{h_{00,ii}}{2}.$$

The result of the gravitational red shift experiment determines h_{00}: from eqn (2.2), for instance,

$$h_{00} = -\frac{2GM}{rc^2} = \frac{2\varphi}{c^2}.$$

where φ is the gravitational potential. If this assignment is carried through here, then

$$R_{00} = \frac{\varphi_{,ii}}{c^2} = \frac{\nabla^2\varphi}{c^2}.$$

Suppose that the rest density of matter is ρ; then

$$T_{00} = \rho c^2.$$

It is a sufficient approximation to take $g_{00} = 1$ on the right-hand side of eqn (7.20), so that

$$T_{00} - \frac{T^0{}_0 g_{00}}{2} = \frac{\rho c^2}{2}.$$

Finally, substituting on both sides in eqn (7.20) gives

$$\nabla^2 \varphi = 4\pi G\rho - \Lambda c^2.$$

This reproduces eqn (7.9), the differential form of Newton's law of gravitation, if Λ vanishes. The comparison verifies that the constant appearing in Einstein's equation has magnitude $8\pi G/c^4$; any other choice would spoil the agreement with Newton's law in the classical limit.

We can take the last equation and use it to calculate the classical force per unit mass due to the cosmological constant. This requires us to reverse the steps which led to eqn (7.9). We obtain the force on unit mass:

$$F = \frac{-GM}{r^2} + \frac{c^2 \Lambda r}{3} \tag{7.21}$$

The experimental limit on Λ given later in Section 11.5 is

$$|\Lambda| \leqslant 10^{-52}\ \mathrm{m}^{-2},$$

so that on Earth the acceleration due to the cosmological term is about 10^{-22} times that due to the gravitational pull of the Sun. Only over cosmologically large distances could the effect of the cosmological term become significant.

At this point the equivalences between classical and general relativistic quantities are collected and reviewed. Making use of the result of the gravitational red shift experiment we have shown that

$$h_{00} = 2\varphi/c^2$$

so that the metric coefficients replace the classical gravitational potential. However, $g_{\mu\nu}$ (and $h_{\mu\nu}$) have six independent components compared with a single classical potential. In the Newtonian limit, $\mathrm{d}x^i/\mathrm{d}\tau \ll \mathrm{d}x^0/\mathrm{d}\tau$ and $\tau \approx t$, so that the geodesic eqn (6.13) reduces to eqn (6.19).

$$\frac{\mathrm{d}^2 x^i}{\mathrm{d}t^2} = -c^2 \Gamma^i_{00},$$

where the Γ^i_{00} are the important metric connections in the Newtonian limit. To first order in h,

$$\Gamma^i_{00} = -\frac{1}{2}\eta^{ii}\frac{\partial h_{00}}{\partial x^i} = \frac{\partial\varphi/\partial x_i}{c^2}$$

(with no summation over i). We can reiterate the conclusion drawn in Section 6.8 that the metric connections replace both the inertial and the gravitational forces of classical mechanics. Finally, eqn (7.5) permits an interpretation of the Riemann curvature tensor in classical terms. In the Newtonian limit all the time derivatives are small compared with spatial derivatives, and so eqn (7.5) reduces to

$$R_{k0j0} = -\frac{1}{2}g_{00,kj} = -\frac{(\partial^2\varphi/\partial x^k \partial x^j)}{c^2}$$

which is a component of the *tidal* force. Table 7.1 summarizes the correspondence between general relativistic and Newtonian kinematic quantities.

Table 7.1 General relativistic quantities and their Newtonian analogues

GR quantities	*Newtonian analogues*
$g_{\mu\nu} = 1 + h_{\mu\nu}$ metric tensor	$1 + 2\varphi/c^2$ gravitational potential φ/c^2
$g_{\mu\nu,\alpha}$, $\Gamma^{\mu}{}_{\nu\alpha}$	$(\partial\varphi/\partial x^{\alpha})/c^2$ gravitational force/c^2
$g_{\mu\nu,\alpha\beta}$, $R^{\mu}_{\nu\alpha\beta}$	$(\partial^2\varphi/\partial x^{\alpha}\partial x^{\beta})/c^2$ tidal force/c^2

8
Tests of general relativity

Since 1960 very precise tests of GR have become possible thanks to advances in technology. The experiments have made use of space probes, atomic clocks, very large radio-telescope arrays, laser-ranging measurements of the Earth–Moon distance and radar-ranging measurements of the distances to nearby planets. Collection and analysis of the data have required fast electronics, high-speed data links and of course powerful computing facilities.

The measurements made on GR effects observed within the solar system are described first: the excess advance of the perihelion of Mercury is described in Section 8.1, the deflection of radiation by the Sun in Section 8.2 and the time delay of radar signals passing by the Sun in Section 8.3. Taken together with the measurement of the gravitational spectral shift these measurements make up the classical tests of GR. More recently, tests have been made to check that orbits calculated according to GR simultaneously fit the planetary orbital data from a wide range of measurements (Section 8.4).

Beyond the solar system the possibilities for tests dwindle away: for example, radar echoes from nearby stars are undetectably weak and anyway return years later! However, Einstein predicted that a galaxy could act as a gravitational lens bending light from a more distant source. Recently, several examples have been discovered where multiple images of a single quasar are seen. This qualitative evidence for a GR effect on the cosmic scale is discussed in Section 8.5. More surprisingly, several effects due to GR are in evidence for the binary pair containing the pulsar PSR 1913+16. How these effects have been unravelled from measurements of the arrival times of the pulsar pulses is recounted in Section 8.6.

8.1 The perihelion advance of Mercury

Mercury, the innermost planet, follows an elliptical orbit at a mean distance of 58 million km from the Sun. Other planets attract Mercury and perturb its orbit so that the long axis of the ellipse slowly rotates in its plane with respect to the frame of the distant galaxies. The point at which the planet is nearest to the Sun is called its perihelion, and so this motion is known as precession of the perihelion. Calculations using Newtonian mechanics predict a precession of 532 arcsec per century. However, the observed precession is 43.11(45) arcsec larger; this notation means 43.11 arcsec with an error of 45 in the last two digits , i.e. an error of ± 0.45 arcsec. The discrepancy was recognized in

1859 by Leverrier and it was proposed that a small undetected planet was the cause. Careful scrutiny by telescopes and space probes reveals no trace of any such object. Venus and the Earth also show smaller unexplained residual advances of their perihelia.

Three centuries of telescopic measurements of the planets have been supplemented by decades of radar echo detection which determine the distance of Mercury and other planets directly (Shapiro *et al.* 1971). Figure 8.1 illustrates recent techniques (Hellings 1984) using one of the 64 m antennae of the Deep Space Network (DSN). A radar carrier of 1 ms duration (300 km long) is phase modulated with a random 255 bit code. When the pulse echo $E(t + t_0)$ returns it is cross-correlated with the pattern transmitted $T(t)$ by forming the sum of products $S(t_0) = \sum_t T(t)\, E(t + t_0)$. The delay t_0 is only known approximately, and so the cross-correlation is repeated for 3 μs steps in t_0 (i.e. every 1000 m). If the echo and pulse patterns match, then the correlator output is large; otherwise, even for a 3μs offset, the output is small. In practice the output rises sharply when the echo returns from the nearest point on the planet but dies away slowly as echoes arrive from surrounding regions of the planet's surface. The delay until the initial sharp rise in output from the correlator gives the distance to the planet. Corrugations of terrain limit the accuracy to ± 1 km.

The analysis of Mercury's motion commences with the statement that Mercury follows a geodesic in the Schwarzschild space–time around the Sun. Perturbations to Mercury's orbit due to other planets can be ignored because these are nearly independent of the GR-induced precession. Orbits in Schwarzschild space–time are of general interest, and so the opportunity will be taken to include a number of comments not directly relevant to the calculation of Mercury's orbit. We assume that the orbital plane has $\theta = \pi/2$ so that the metric eqn (4.10) becomes

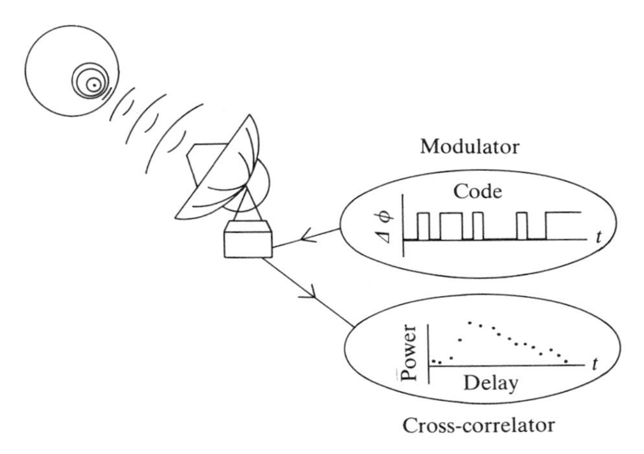

Fig. 8.1 The use of radar ranging to measure planetary distances. (After Hellings 1984.)

$$c^2\,\mathrm{d}\tau^2 = c^2 Z\,\mathrm{d}t^2 - \mathrm{d}r^2/Z - r^2\,\mathrm{d}\varphi^2, \tag{8.1}$$

where $Z = (1 - 2GM/rc^2)$, M is the Sun's mass, and r is the distance of Mercury from the Sun. Then $Z \approx 1 - 5 \times 10^{-7}$. Dividing eqn (8.1) by $\mathrm{d}\tau^2$ and then multiplying by the square of the planet's mass (m^2) gives

$$m^2c^2 = m^2c^2 Z\left(\frac{\mathrm{d}t}{\mathrm{d}\tau}\right)^2 - \frac{m^2(\mathrm{d}r/\mathrm{d}\tau)^2}{Z} - m^2r^2\left(\frac{\mathrm{d}\varphi}{\mathrm{d}\tau}\right)^2. \tag{8.2}$$

In flat space–time eqn (8.2) reduces to

$$m^2c^2 = m^2c^2\left(\frac{\mathrm{d}t}{\mathrm{d}\tau}\right)^2 - m^2\left(\frac{\mathrm{d}r}{\mathrm{d}\tau}\right)^2 - m^2\,r^2\left(\frac{\mathrm{d}\varphi}{\mathrm{d}\tau}\right)^2,$$

i.e. to

$$m^2c^2 = m^2c^2\gamma^2 - m^2{v_r}^2\gamma^2 - m^2{v_\varphi}^2\gamma^2,$$

where v_r and v_φ are the radial and tangential components of the velocity v and $\gamma = 1/(1 - v^2/c^2)^{1/2}$. This last equation is the standard SR formula relating rest mass to four-momentum:

$$m^2c^2 = E^2/c^2 - p^2$$

and eqn (8.2) is its equivalent in Schwarzschild space–time. The geodesic equation for Mercury in its integral form is given by eqn (6.18):

$$\delta\int \mathrm{d}\tau\left[Zc^2\left(\frac{\mathrm{d}t}{\mathrm{d}\tau}\right)^2 - \frac{(\mathrm{d}r/\mathrm{d}\tau)^2}{Z} - r^2\left(\frac{\mathrm{d}\varphi}{\mathrm{d}\tau}\right)^2\right] = 0.$$

General solutions for such equations are discussed in Appendix A. In particular if the integrand L in square brackets is independent of a coordinate x^μ, then the quantity $\partial L/\partial(\mathrm{d}x^\mu/\mathrm{d}\tau)$ is conserved (A.3). Here the integrand is independent of t so that one conservation law is

$$\frac{\partial L}{\partial(c\,\mathrm{d}t/\mathrm{d}\tau)} = \text{constant},$$

so that

$$2Zc\frac{\mathrm{d}t}{\mathrm{d}\tau} = \text{constant}. \tag{8.3}$$

Re-expressing this result in terms of the momentum component p_0, we have

$$cp_0 = Zcp^0 = Zmc^2\frac{\mathrm{d}t}{\mathrm{d}\tau} = E,$$

where E is a constant. Thus for a body in free fall E is a constant of motion, and in the absence of gravitational forces E reduces to $mc^2\gamma$, which is the usual SR energy. In SR E would be constant in the absence of any forces, while in

Schwarzschild space–time E is constant for a body in free fall. The integrand given above is also independent of φ, so that

$$\frac{\partial L}{\partial(\mathrm{d}\varphi/\mathrm{d}\tau)} = \text{constant}$$

i.e.

$$r^2\frac{\mathrm{d}\varphi}{\mathrm{d}\tau} = J, \quad \text{a constant.} \tag{8.4}$$

Equation (8.4) is the equivalent of the classical law of conservation of angular momentum for Schwarzschild space–time. Replacing $\mathrm{d}t/\mathrm{d}\tau$ in eqn (8.2) by E/Zmc^2 gives

$$\frac{E^2}{Zc^2} - \frac{m^2(\mathrm{d}r/\mathrm{d}\tau)^2}{Z} - m^2r^2\left(\frac{\mathrm{d}\varphi}{\mathrm{d}\tau}\right)^2 = m^2c^2. \tag{8.5}$$

Multiplying this by Z and dropping a factor m throughout we obtain

$$\frac{E^2}{mc^2} - m\left(\frac{\mathrm{d}r}{\mathrm{d}\tau}\right)^2 - Zr^2m\left(\frac{\mathrm{d}\varphi}{\mathrm{d}\tau}\right)^2 = mc^2 - \frac{2GMm}{r},$$

which can be rearranged as

$$\frac{m(\mathrm{d}r/\mathrm{d}\tau)^2}{2} + \frac{mr^2(\mathrm{d}\varphi/\mathrm{d}\tau)^2Z}{2} - \frac{GMm}{r} = \frac{(E^2/mc^2 - mc^2)}{2} = T \tag{8.6}$$

where T is also a conserved quantity. Equation (8.6) is the equivalent of the classical conservation of energy equation for Schwarzschild space–time. There is a 'radial kinetic energy term' $m(\mathrm{d}r/\mathrm{d}\tau)^2/2$, a 'transverse kinetic energy term' $mr^2(\mathrm{d}\varphi/\mathrm{d}\tau)^2Z/2$, and a gravitational energy term $-GMm/r$. Together, eqns (8.3), (8.4), and (8.6) fully describe the motion of Mercury or for that matter any test mass in free fall in Schwarzschild space–time. The quantities $E = cp_0$ and $J = r^2(\mathrm{d}\varphi/\mathrm{d}\tau)$ are invariants of the motion. We next go on to solve these equations of motion. Using eqn (8.4) gives

$$\frac{\mathrm{d}r}{\mathrm{d}\tau} = \frac{\mathrm{d}r}{\mathrm{d}\varphi}\frac{\mathrm{d}\varphi}{\mathrm{d}\tau} = \frac{J}{r^2}\frac{\mathrm{d}r}{\mathrm{d}\varphi},$$

and putting $u = 1/r$ this becomes

$$\frac{\mathrm{d}r}{\mathrm{d}\tau} = -J\frac{\mathrm{d}u}{\mathrm{d}\varphi}.$$

Substituting this expression for $\mathrm{d}r/\mathrm{d}\tau$ into eqn (8.6) gives

$$\frac{J^2(\mathrm{d}u/\mathrm{d}\varphi)^2}{2} + \frac{u^2J^2Z}{2} - GMu = \frac{T}{m}.$$

Differentiating this with respect to φ and cancelling a factor $\mathrm{d}u/\mathrm{d}\varphi$, we obtain

$$J^2\left(\frac{\mathrm{d}^2u}{\mathrm{d}\varphi^2}\right) + J^2u - \frac{3GMu^2J^2}{c^2} - GM = 0.$$

Rearrangement gives

$$\frac{\mathrm{d}^2u}{\mathrm{d}\varphi^2} + u - \frac{GM}{J^2} = \frac{3GMu^2}{c^2}. \tag{8.7}$$

This result can be compared with the Newtonian equation for orbits in the gravitation potential of a mass M:

$$\frac{\mathrm{d}^2u}{\mathrm{d}\varphi^2} + u - \frac{GM}{J^2} = 0. \tag{8.7N}$$

The solution of eqn (8.7N) is well known to be

$$u = \frac{1 + e\cos\varphi}{l}$$

where $l = a(1 - e^2)$. Figure 8.2 shows a bound orbit: an ellipse with eccentricity $0 \leqslant e < 1$. At aphelion $\varphi = \pi$, $r = a(1 + e)$; at perihelion $\varphi = 0$, $r = a(1 - e)$. Hence the long (major) axis is $2a$ in length. In addition although less easy to prove,

$$l = J^2/GM.$$

A solution of eqn (8.7N) is clearly a very good approximate solution of eqn (8.7) because Mercury's orbit is nearly Newtonian. Consequently we can rewrite the small term on the right-hand side of eqn (8.7) as

$$3GM(1 + e\cos\varphi)^2/l^2c^2$$

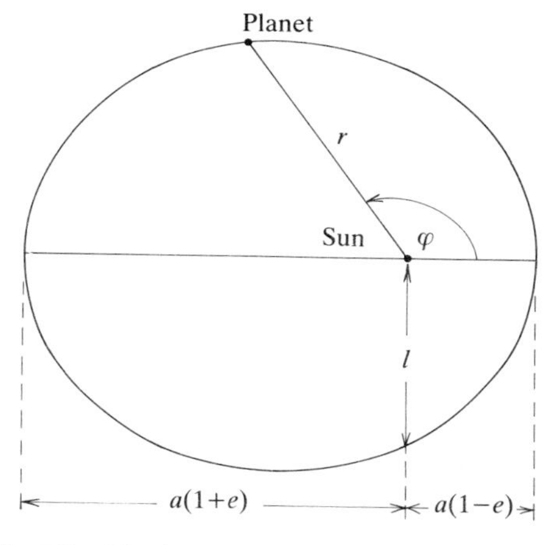

Fig. 8.2 The bound elliptical orbit of a planet.

and make an entirely negligible error. With this substitution eqn (8.7) becomes

$$\frac{d^2u}{d\varphi^2} + u - \frac{GM}{J^2} = \frac{3GM}{l^2c^2}(1 + 2e\cos\varphi + e^2\cos^2\varphi) \tag{8.7$'$}$$

The solution of eqn (8.7′) is similar to that for (8.7N) with extra particular integral terms arising from the three terms on the right-hand side. Explicitly

$$u = \frac{(1 + e\cos\varphi)}{l} + \frac{3GM}{l^2c^2}\left[\left(1 + \frac{e^2}{2}\right) - \left(\frac{e^2}{6}\right)\cos 2\varphi + e\varphi\sin\varphi\right]$$

Of the additional terms, the first is a constant and the second oscillates through two cycles on each orbit; both these terms are immeasurably small. However, the last term increases steadily in amplitude with φ, and hence with time, whilst oscillating through one cycle per orbit, and so this term is responsible for the precession. Dropping the unimportant terms we have

$$u = \frac{1 + e\cos\varphi + e\alpha\varphi\sin\varphi}{l}$$

where the factor $\alpha = 3GM/lc^2$ is extremely small. Thus

$$u = \frac{1 + e\cos[(1 - \alpha)\varphi]}{l} \tag{8.8}$$

is the general relativistic solution. At perihelion we now have

$$(1 - \alpha)\varphi = 2n\pi,$$

i.e.

$$\varphi = 2n\pi + 6n\pi\frac{GM}{lc^2}$$

where n is an integer. This shows that the perihelion advances by $\Delta\varphi = 6\pi GM/lc^2$ per rotation and so the rate of precession is

$$\frac{\Delta\varphi}{\tau} = \frac{6\pi GM}{a(1 - e^2)\tau c^2}. \tag{8.9}$$

Table 8.1 compares the observed with the predicted planetary precessions, measured in arcseconds per century. There is excellent agreement. Examination of eqn (8.9) makes it clear why Mercury shows the largest effect. Mercury is closer to the Sun (small a), and has a shorter period and a larger orbital eccentricity (small $(1 - e^2)$). Moreover the greater the eccentricity the more precisely can the passage through perihelion be timed; conversely with a circular orbit ($e = 0$) the precession becomes undetectable!

It is worth noting that if the Sun were an oblate spheroid (squashed at its poles), this would also have the effect of causing the orbit of Mercury to precess. Measurements on the shape of the Sun are difficult to interpret (see

Table 8.1 Comparison of observed with predicted planetary precessions

Planet	*Observed precession* (arcsec per century)	*Predicted precession* (arcsec per century)
Mercury	43.11 ± 0.45	43.03
Venus	8.4 ± 4.8	8.6
Earth	5.0 ± 1.0	3.8

Dicke 1983), but it appears that the taking account of the small oblateness does not spoil the agreement between the observed precession and that calculated according to GR (see Claverie *et al.* 1981).

8.2 The deflection of light by the Sun

Measurements of this deviation were discussed in Chapter 2. The calculation of the orbit of a photon in Schwarzschild space around the Sun follows the same steps as that for material particles given in Section 8.1. For a photon the left-hand side of eqn (8.1) is zero and the equivalent of eqn (8.7) is

$$\frac{d^2u}{d\varphi^2} + u = \frac{3GMu^2}{c^2} \qquad (8.10)$$

for light rays travelling in the equatorial plane. The approximate solution of this equation, when the small term on the right-hand side is neglected, is

$$u = \frac{\cos(\varphi + \alpha)}{b}$$

where b and α are constants of integration. This is the equation of a straight line with distance b of closest approach to the Sun. Usually b is called the impact parameter. A simple rotation of the axis defining $\varphi = 0$ can be used to eliminate α (see Fig. 8.3):

$$u = \cos\varphi/b. \qquad (8.11)$$

Then, following the approach of Section 8.1, we substitute this value for u on the right-hand side of (8.10) to give

$$\frac{d^2u}{d\varphi^2} + u = \frac{3GM\cos^2\varphi}{b^2c^2}. \qquad (8.12)$$

Equation (8.12) has a particular integral

$$u = \frac{GM(2 - \cos^2\varphi)}{b^2c^2}$$

and its full solution is

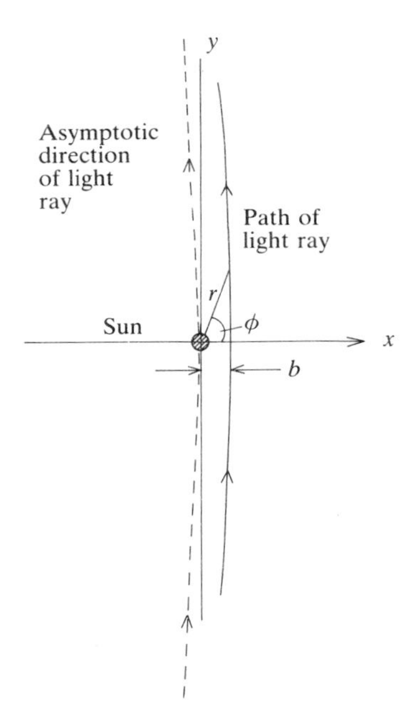

Fig. 8.3 The deflection of a light ray passing near to the Sun.

$$u = \frac{\cos \varphi}{b} + \frac{GM(2 - \cos^2 \varphi)}{b^2c^2} \tag{8.13}$$

At distant points $u \rightarrow 0$ and eqn (8.13) becomes

$$\frac{\cos \varphi}{b} + \frac{2GM}{b^2c^2} - \frac{GM \cos^2 \varphi}{b^2c^2} = 0.$$

Also $\cos \varphi \rightarrow 0$ at distant points; therefore the $\cos^2 \varphi$ term can be omitted, giving

$$\cos \varphi = -2GM/bc^2.$$

As a result φ tends asymptotically to $\pi/2 + 2GM/bc^2$ in one direction and to $3\pi/2 - 2GM/bc^2$ in the other. This makes the total deflection

$$\Delta\varphi = 4GM/bc^2 \tag{8.14}$$

which for light just grazing the Sun's limb ($b = b_0 = 6.96 \times 10^8$ m) is 1.750 arcsec. At any larger impact parameter b, the deflection $\Delta\varphi$ is scaled down by b_0/b. This prediction of GR is confirmed by both the optical and radio measurements discussed in Chapter 2. If the same calculation is made taking into account only the time distortion implied by the equivalence principle, the resultant deviation is only half as large. If the contribution of frame and time distortion are to be separated explicitly, the calculation of the deflection of light becomes more involved. This instructive calculation is presented in Appendix G.

8.3 Radar echo delays

Along the path of a light ray travelling in the equatorial plane in Schwarzschild space eqn (8.1) reduces to

$$0 = c^2 Z\,\mathrm{d}t^2 - \mathrm{d}r^2/Z - r^2\,\mathrm{d}\varphi^2 \tag{8.15}$$

where Z is $1 - 2GM/rc^2$. The tangential and radial coordinate velocities are obtained by setting dr and dφ to zero in turn. Then

$$r\frac{\mathrm{d}\varphi}{\mathrm{d}t} = c\sqrt{Z} \qquad \text{and} \qquad \frac{\mathrm{d}r}{\mathrm{d}t} = cZ,$$

showing that as light approaches the origin its coordinate velocity falls. The delays in travel time of radar reflections from the planets have been measured by Shapiro *et al.* (1971) as the planets pass behind the Sun (i.e. at superior conjunction). Figure 8.4 shows a radar beam travelling from Earth (E) to Venus (V) and passing a distance b from the sun. Using eqns (8.3) and (8.4) we can eliminate φ from eqn (8.15):

$$r^2\left(\frac{\mathrm{d}\varphi}{\mathrm{d}t}\right) = \frac{r^2(\mathrm{d}\varphi/\mathrm{d}\tau)}{\mathrm{d}t/\mathrm{d}\tau}$$

$$= Z\sqrt{W}$$

where W is a constant of motion. Then eqn (8.15) becomes

$$0 = Zc^2 - \frac{(\mathrm{d}r/\mathrm{d}t)^2}{Z} - \frac{WZ^2}{r^2}.$$

At the point nearest to the Sun, $\mathrm{d}r/\mathrm{d}t = 0$, $r = b$, and $Z = Z_b = 1 - 2GM/bc^2$, so that this equation reduces to

$$W = b^2c^2/Z_b.$$

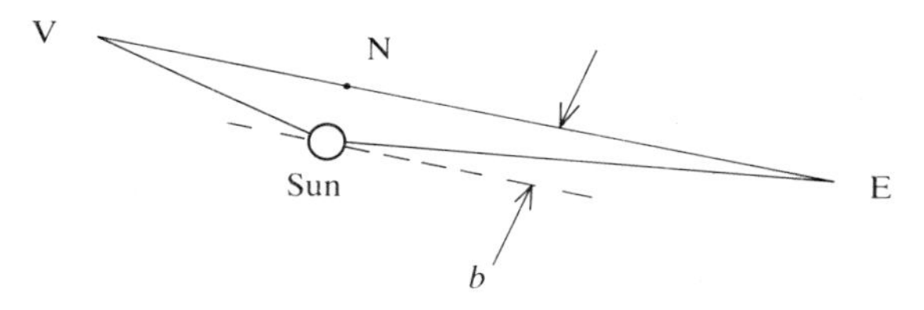

Fig. 8.4 A radar beam from Earth to Venus near superior conjunction.

With this value substituted for W in the previous equation we have

$$\frac{\mathrm{d}r}{\mathrm{d}t} = cZ\left(1 - \frac{b^2 Z}{r^2 Z_b}\right)^{1/2}. \tag{8.16}$$

Integrating eqn (8.16) gives the travel time from Earth to N as

$$t_{\mathrm{E}} = \int_{\mathrm{N}}^{\mathrm{E}} \frac{\mathrm{d}r}{cZ(1 - b^2 Z/r^2 Z_b)^{1/2}}.$$

The integrand can be expanded to first order in GM/rc^2:

$$t_{\mathrm{E}} = \int_{\mathrm{N}}^{\mathrm{E}} \left[\frac{r\,\mathrm{d}r}{c(r^2 - b^2)^{1/2}}\right]\left[1 + \frac{2GM}{rc^2} + \frac{GMb}{r(r+b)c^2}\right].$$

Then

$$t_{\mathrm{E}} = \frac{(r_{\mathrm{E}}^2 - b^2)^{1/2}}{c} + \frac{2GM}{c^3}\ln\left[\frac{r_{\mathrm{E}} + (r_{\mathrm{E}}^2 - b^2)^{1/2}}{b}\right] + \frac{GM}{c^3}\left(\frac{r_{\mathrm{E}} - b}{r_{\mathrm{E}} + b}\right)^{1/2}.$$

The first component of t_{E} is the journey time in flat space–time; hence the excess time taken is

$$\Delta t_{\mathrm{E}} \approx \left(\frac{2GM}{c^3}\right)\ln\left[\frac{r_{\mathrm{E}} + (r_{\mathrm{E}}^2 - b^2)^{1/2}}{b}\right] + \frac{GM}{c^3}.$$

There is a similar expression for the time delay from N to Venus. Adding this in and multiplying by a factor of 2 produces the delay for the round trip Earth–Venus–Earth:

$$\Delta t \approx \frac{4GM}{c^3}\left(\ln\left\{[r_{\mathrm{E}} + (r_{\mathrm{E}}^2 + b^2)^{1/2}]\frac{r_v + (r_v^2 + b^2)^{1/2}}{b^2}\right\} + 1\right)$$

$$\approx \frac{4GM}{c^3}\left[\ln\left(\frac{4r_{\mathrm{E}}r_v}{b^2}\right) + 1\right]. \tag{8.17}$$

A round trip to Venus requires 1300 s at superior conjunction while the delay predicted with eqn (8.17) is only 220 μs. Figure 8.5 shows measurements of over 600 days of the radar time delay for reflections from Venus (Shapiro *et al.* 1971) using a radio-telescope at the Haystack Observatory, Massachusetts, at 7.84 GHz and one at Arecibo, Puerto Rico, at 430 MHz. The excess time delay is the difference between the observed travel time and that calculated in flat space–time for the observed orbits of Venus and the earth. The solid curve in Fig. 8.5 is the GR prediction and fits the changing delay as Venus moves through superior conjunction extremely well. One experimental difficulty is that the solar corona has a finite refractive index (varying as $1/(\text{frequency})^2$) and the effect of the corona is to increase the delay. A correction for this effect has already been included in the comparison made in Fig. 8.5. A second difficulty lies in the uncertainty in our knowledge of the topography of Venus; the precision is only ± 1500 m which leads to an uncertainty of 10 μs in timing.

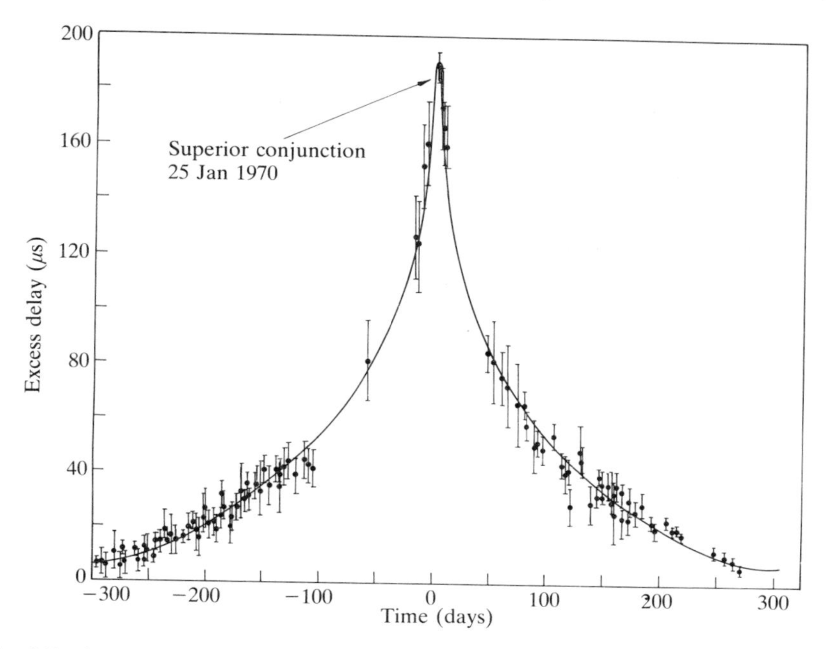

Fig. 8.5 A sample of post-fit residuals for Earth–Venus time-delay measurements: ———, prediction using GR (Shapiro *et al.* 1972). (Courtesy Professor Shapiro and *Physical Review Letters*, published by the American Physical Society.)

Both these difficulties were overcome in a later experiment (Reasenberg *et al.* 1979) to measure the time delay for reflections from Mars over a period of 14 months. First of all the radar signal was received and retransmitted by the Viking Lander sitting on the Martian surface. This eliminated the uncertainty due to topography. Secondly, the effect of the solar corona was determined by studying the delay at two frequencies (2.3 and 8.4 GHz) by means of other transponders on a Viking Orbiter spacecraft in Mars orbit. Knowing the way that the refractive index of the solar corona depends on frequency ($1/f^2$), it was possible to determine the magnitude of the effect from the delays measured at the two frequencies. Timing uncertainties were reduced to below about 0.1 μs. The ratio found between the delay observed and the delay expected according to the general theory was 1.000 ± 0.002, which is another clear success for Einstein's theory.

8.4 An overall test in the solar system

A recent study by Hellings (1984) goes beyond the test of separate GR predictions in isolation. He describes how a large body of data concerning planetary motion has been used to make an overall test. For this test a parametrization of the metric is chosen which is general enough to include deviations from Newtonian mechanics which are wider than general relativistic effects (this post-Newtonian parametrization has also been applied to analysing individual effects). The overall fit tests how unique the explanation provided by GR for the observed departures from classical behaviour may be. The data sample contained the following:

(1) 44000 measurements of transits across the meridian (at Washington) of the Sun, Moon and planets (accuracy ± 1 arcsec);
(2) 3000 laser measurements to a corner-cube reflector placed on the Moon by Apollo astronauts (accuracy ± 0.1 m);
(3) 3000 radar range measurements to Venus and Mercury (± 1 km) and to Mars (part ± 100 m; part ± 7 m);

Orbits were fitted to this data set using a metric with

$$g_{00} = 1 - \frac{2GM}{r} + \frac{2\beta G^2 M^2}{c^2 r^2} + \alpha_1 \frac{GM}{r}\left(\frac{w}{c}\right)^2$$

$$g_{0k} = \alpha_1 \left(\frac{GM}{r}\right)\left(\frac{w^k}{c}\right)$$

$$g_{ij} = -\left(1 + \frac{2\gamma GM}{r}\right)\delta_{ij}$$

for isotropic coordinates. w^k is the velocity of the Earth relative to a preferred frame defined by the cosmic background radiation (see Chapter 11). In GR α_1, β, and γ would be exactly 0, 1, and 1 respectively. Hellings' fit yielded $\alpha_1 = (2.2 \pm 1.8) \times 10^{-4}$, $(\beta - 1) = (0.2 \pm 1.0) \times 10^{-3}$, and $(\gamma - 1) = (-1.2 \pm 1.6) \times 10^{-3}$, which confirms the unique consistency of GR with measurements in the solar system. The reader can refer to Will (1981) for more details concerning the post-Newtonian parametrization of solar system data.

8.5 The discovery of gravitational lenses

Einstein pointed out that if a galaxy lay between the Earth and a distant source of light, the light would be focused by the gravitational field of the galaxy. Two or more images would then be seen, with cases of high symmetry producing a ring image. Recent studies of quasars have identified multiple images produced in this way. Quasars are very luminous sources of small angular size which were first discovered with radio-telescopes. Their optical spectra show large red shifts z, where

$$z = \frac{\lambda_0 - \lambda_E}{\lambda_E},$$

λ_E being the wavelength of a spectral line from a source on Earth and λ_0 the wavelength of the same line observed in the quasar spectrum. This shift is proportional to the distance d from Earth:

$$z = Hd/c$$

where H is the Hubble constant and is a measure of the rate of expansion of the Universe. At present, $H = H_0 = 75 \pm 25$ km s^{-1} Mpc^{-1}, and has probably fallen since the creation of the Universe (see Chapter 11). Quasars have red shifts ranging up to 4.5, making them the most distant distinct sources so far detected. Quasars are widely believed to be the cores of young galaxies. Several pairs of quasars have been detected which have virtually identical red shifts and are separated by small angular intervals: these pairs are now interpreted as twin images of a single quasar. One case reported by Surdej *et al.* (1987) provides unequivocal evidence for this interpretation. The quasar pair UM673A and UM673B both have red shifts of 2.72 and are separated by 2.2 arcsec. Figure 8.6 shows the spectra of UM673A and UM673B; these are evidently identical in all features, with one being somewhat brighter. The red shifts of these spectra give values of zc which differ by only 24 ± 109 km s^{-1} in 281 600 km s^{-1}. It seems inescapable that two images of a single quasar have been observed. Many similarly convincing examples have been reported recently. The experimental evidence of gravitational lensing is strong qualitative evidence that GR applies over the whole of the Universe and not just locally in our solar system.

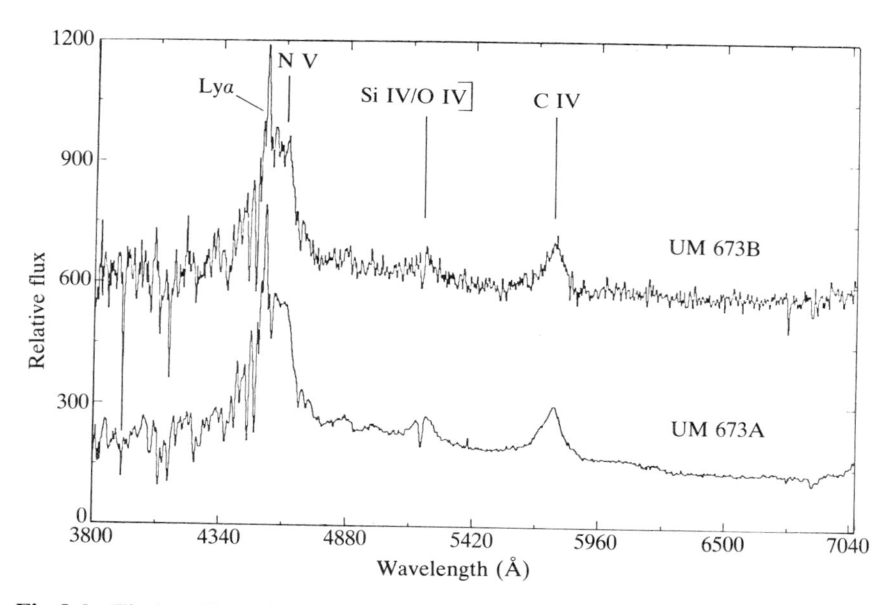

Fig. 8.6 The low dispersion spectra of UM 673A and UM673B recorded in December 1986. The resolution is about 1.3 nm. (With kind permission of Nature; courtesy J. Surdej.)

8.6 PSR 1913+16

The pulsar PSR 1913+16 is one member of a binary pair located 5 kpc from the Earth in the constellation Aquila. Both the pulsar and its companion have masses close to 1.4 $M_\odot$ and move in tight orbits which would almost fit inside the Sun ($M_\odot$ is one solar mass). For such a configuration the parameter GM/rc^2 is large, and therefore general relativistic effects are expected to be important. Indeed precision measurements of three general relativistic effects have been possible: namely, the orbital precession, the gravitational red shift, and the decay of the orbits due to gravitational radiation. It is remarkable that such detailed information can be extracted for such a remote system. This information is only retrievable because nature has provided an extremely precise clock on one star, namely the pulsar itself.

Pulsars are compact stars a few kilometres in diameter with masses around 1.4 $M_\odot$. They are the final active phase in the lives of many intermediate mass stars: the core has by then contracted to a neutron star and in the same process the outer layers of the star have rebounded as a supernova explosion. Contraction leads to a compression of the magnetic field which can then attain very high values ($\sim 10^8$ T). The pulsar is also spinning rapidly with a period between 1 ms and a few seconds, but its magnetic dipole axis is not aligned with the spin axis. By some plasma mechanism intense coherent beams of radiation are produced along the magnetic axis; these beams swing round like searchlights as the neutron star rotates. A beam from a favourably located pulsar can illuminate the Earth once per cycle and this behaviour is detected, usually with radio-telescopes, as pulses of radiation at regular intervals. The stability of isolated pulsars rivals that of atomic clocks; for example PSR 1937+21 has a period P of 1.5578 ms with a rate of change of only 10^{-19} s s^{-1}. It is this property of long-term frequency stability that makes precision measurements on general relativistic effects possible in the case of PSR 1913+16.

PSR 1913+16 was discovered by Hulse and Taylor (1975) using the 305 m radio-telescope at Arecibo. The signal is a regular train of radio-frequency pulses with a spacing between pulses of 59 ms. Hulse and Taylor found that the period changed by as much as 80 μs with a 7.75 h cycle, which is uncharacteristic behaviour for a pulsar. They deduced that the pulsar must be in orbit round a companion with the result that the pulse frequency was being Doppler shifted by the pulsar's motion in orbit. This fixes the *orbital* period to be 7.75 hs. Unless the companion is vastly more massive than the Sun, this also indicates that the stars have small orbits. Neither optical nor radio signals have been detected from the companion, and so it is also likely to be a compact star. However, it cannot be very much lighter than the pulsar, otherwise the pulsar motion would not produce such a large Doppler shift.

Figure 8.7(a) shows the variation of the line-of-sight velocity of PSR 1913+

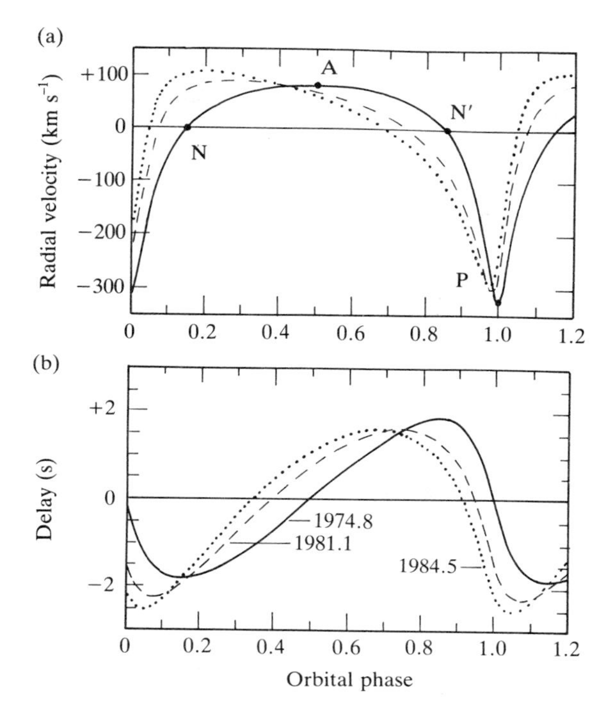

Fig. 8.7 (a) The variation of the line-of-sight velocity component of the pulsar PSR 1913+16 over complete cycles for three epochs; (b) the time delay as a function of orbital phase for the same epochs as for Fig. 8.7(a). (Courtesy Professor Taylor and Cambridge University Press (1986).)

16 over a complete orbit at three different epochs. At epoch 1974.8 the curve is symmetric about the mid-point; hence the orbit must be symmetric with respect to the line of sight $\boldsymbol{n}$. The inferred orientation of the orbit is shown in Figs. 8.8(a) and 8.8(b) with the major axis AP lying perpendicular to $\boldsymbol{n}$ and the orbital plane tilted with respect to $\boldsymbol{n}$. O is the centre of mass of the binary. The portion of its orbit over which the pulsar approaches the Earth (negative radial velocity) is indicated by heavier lines in Fig. 8.8(b). Corresponding points on Figs. 8.7(a) and 8.8(b) are labelled by the same letter (P, N, A, and N′). Another simple deduction from Fig. 8.7(a) is that the orbit is quite eccentric; had the orbit been circular, the velocity curve would be sinusoidal with a mean radial velocity of zero. In summary the binary consists of two compact stars of not dissimilar mass moving in tight eccentric orbits. Reference back to Section 8.1 reminds us that these conditions will maximize the general relativistic precession of the pulsar's orbit.

The angular velocity of precession of the periastron (point of nearest approach of the pulsar and its companion) is given by eqn (8.9):

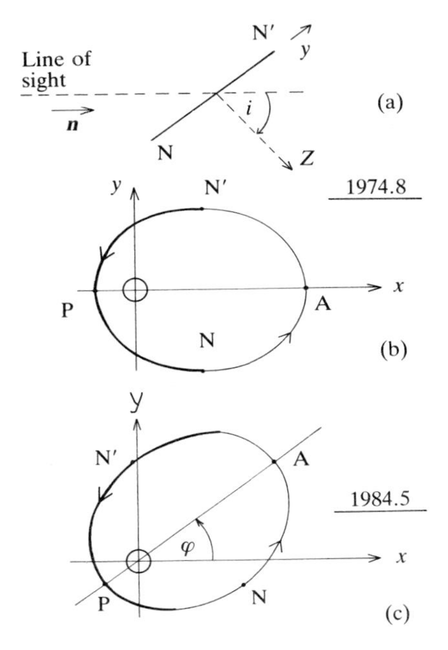

Fig. 8.8 The orbit of PSR 1913+16: (a) side view showing the inclination i of the orbital plane to the line of sight $\boldsymbol{n}$ from the Earth; (b) a plan view at epoch 1974.8 when the major axis AP was perpendicular to the line of sight; (c) a plan view at epoch 1984.5 with Ox perpendicular to the line of sight.

$$\dot{\varphi} = \frac{6\pi GM'}{a(1 - e^2)\tau c^2}$$

where a is the semi-major axis length, e is the eccentricity, τ is the orbital period, and M' is the mass of the companion. Comparing the pulsar orbit with the orbit of Mercury; a, $1 - e^2$, and τ are now all smaller, and so the precession ought to be much more rapid for PSR 1913+16. The effect is plain to see in the way that the velocity curves change with time in Fig. 8.7(a). By the epoch 1981.1 the symmetry seen at 1974.8 is lost, which indicates that the orbit has turned substantially. Figure 8.8(c) shows its orientation at epoch 1984.5, with a periastron advance of over 40° since 1974.8.

Pulse arrival times have been recorded by Taylor and his colleagues over extended periods during the last 15 years. The accuracy with which the pulses can be timed is such that the phase (Fig. 8.7(b)) can be carried across gaps in the data with an accuracy to better than 20 μs; that is to better than $\{(20 \times 10^{-6})/(59 \times 10^{-3})\}2\pi \approx 2 \times 10^{-3}$ rad. Weissberg and Taylor (1984) have summarized the results of a full analysis of the data to that time. A simplified account will be given here. More details can be obtained from Shapiro and Teukolsky (1983).

In the Newtonian approximation the rate $\mathrm{d}A/\mathrm{d}t$ at which the radius vector (from O to the pulsar) marks out the area of the orbit is a constant (see Fig. 8.8(b)). At apastron (A), when the stars are furthest apart,

$$\frac{\mathrm{d}A}{\mathrm{d}t} = \frac{v_a(1+e)a}{2}$$

where v_a is the pulsar velocity and $a(1+e)$ its distance from O. Similarly at periastron (P)

$$\frac{\mathrm{d}A}{\mathrm{d}t} = v_p\frac{(1-e)a}{2}.$$

Taking the ratio of the last two equations, we obtain

$$\frac{v_p}{v_a} = \frac{1+e}{1-e}.$$

Measurements of the Doppler shift at periastron and apastron only give the line of sight components of v_p and v_a. However, their ratio is exactly v_p/v_a, so that the measured ratio yields the eccentricity. Measurement of all pulse arrival times makes it possible to calculate the change in pulse delay due to the change in the distance of PSR 1913+16 from Earth. This delay is shown in Fig. 8.7(b) and reveals that the orbit is about 4 light-seconds across. The path difference across the orbit in Fig. 8.8(a) is (NN′) sin i or $a(1-e^2)^{1/2}\sin i$. e is already determined, and so the delay measurements yield $a \sin i$. Thus far the full analysis gives

$$\tau = 27906.98163(2)s \approx 7.75\ \mathrm{h}$$

$$a \sin i = 2.34185(12)\ \text{light-seconds}$$

$$e = 0.617127(3)$$

$$\varphi = 4.2263(3)^\circ\ \text{per year.}$$

Now, applying Kepler's third law to the pulsar orbit, we obtain

$$\frac{G(M+M')}{(a+a')^3} = \left(\frac{2\pi}{\tau}\right)^2$$

where a' is the semi-major axis of the partner's orbit and M' is its mass; M is the pulsar mass. Also we have

$$a + a' = \frac{(M+M')a}{M'}$$

which can be used to replace $a + a'$ in the previous equation:

$$\frac{a^3(2\pi/\tau)^2}{G} = \frac{(M')^3}{(M+M')^2}.$$

Multiplying both sides by $\sin^3 i$ gives

$$(a \sin i)^3 \frac{(2\pi/\tau)^2}{G} = \frac{(M' \sin i)^3}{(M + M')^2}, \tag{8.18}$$

which is a quantity known to astronomers as the mass function. Equation (8.18) shows that the measurements of τ and $a \sin i$ determine one function of the three physically interesting quantities M, M', and $\sin i$.

The next step in the analysis is to exploit a second general relativistic effect. Note that the orbit is only as large as the Sun's *diameter* and in addition the separation r of the stars varies by a factor $(1 + e)/(1 - e) \approx 4$ over any orbit. There should therefore be a significant and fluctuating gravitational red shift of the pulsar period P:

$$P = P_0 \left(1 - 2\frac{GM'}{rc^2}\right)^{-1}$$

which numerically has an amplitude of order $GM'/ac^2 \approx 10^{-6}$. In addition the pulsar velocity v is so large that the time dilation of SR must be taken into account. When this is done the last equation becomes

$$P = P_0 \left(1 - 2\frac{GM'}{rc^2}\right)^{-1} \left(1 - \frac{v^2}{c^2}\right)^{-1/2}.$$

From Fig. 8.7(a) the variation of v/c over one orbit is seen to have an amplitude of order 10^{-3}, and so the SR time dilation of the period is also of order 10^{-6}. Together the SR and GR effects should lead to a time delay variation of amplitude of about 4 ms over one orbit. This expectation is confirmed: the purely Newtonian analysis leaves an unexplained residual cyclic variation in the delay of the expected magnitude. Here then is confirmation of the gravitational red shift in a sytem 5 kpc from Earth. In the full analysis the measurements of $a \sin i$, τ, φ, and the gravitational red shift suffice to determine M, M', a, and $\sin i$ uniquely. The results of this analysis are

$$M = 1.42(3)M_\odot \qquad M' = 1.40(3)M_\odot \qquad \sin i = 0.76(14).$$

It is worth emphasizing that a fit is made to pulse measurements extending over more than a decade, i.e. over 3×10^8 s, with an accuracy to 10^{-4} s or better! The results, which are in excellent agreement with GR, lend considerable weight to the argument that Einstein's theory applies universally and not just locally.

A potential complication of the analysis could have come from the influence of the partner on the pulsar, had it been comparable with our Sun in size as well as in mass. Intense searches have revealed no optical or other signal from the partner, nor does it eclipse the quasar. It is therefore another compact star, perhaps even a pulsar. The full analysis produces one further surprise. When the orbital period is plotted as a function of time it shows a steady decrease:

$$\frac{d\tau}{dt} = -2.40(9) \times 10^{-12}\ \mathrm{s\,s^{-1}}.$$

This is small but exceedingly well measured. According to Newtonian mechanics the orbital decay for an isolated binary consisting of compact stars is expected to be immeasurably small. In contrast, GR predicts that any such binary should radiate energy continuously in the form of gravitational waves. The predicted energy flux of gravitational radiation (calculated in Chapter 10) leads to a decay rate for the orbit of PSR 1913+16 of

$$\frac{d\tau}{dt} = -2.403(2) \times 10^{-12}\ \mathrm{s\,s^{-1}}.$$

The excellent agreement between the predicted and observed orbital decay rates has a double significance. It offers further confirmation for GR outside the solar system and it provides the first evidence, albeit indirect, for gravitational radiation. This will be discussed further in Chapter 10.

9
Black holes

The general theory of relativity provides a very precise and well-tested description of space–time within the solar system–a region, however, where space–time is nearly flat, with the parameter $2GM/rc^2$ reaching a value of only 10^{-5} at the Sun's surface. At the opposite end of the scale, when $2GM/rc^2$ approaches unity, general relativistic effects are dominant. Such a regime may hold at the centre of galaxies or close to collapsed stars. Because the strong curvature isolates these regions from the rest of the Universe it is not possible to observe them directly.

The Rev. J. Michell in 1784 was the first to notice the implications of the gravitational potential GM/r becoming numerically large (Michell 1784). We write the energy condition for a body of mass m starting with velocity v to escape to infinity from the surface of a star of mass M and radius r. It requires the initial kinetic energy to exceed the gravitational binding energy:

$$mv^2/2 \geqslant GMm/r.$$

i.e.

$$v \geqslant (2GM/r)^{1/2}.$$

Escape is only possible for velocities greater than $(2GM/r)^{1/2}$. As the radius of the star is reduced the escape velocity increases until eventually it reaches the velocity of light. At this stage the radius is

$$r_0 \equiv 2GM/c^2.$$

Michell argued that not even light would be able to escape from any more compact star, and that such a star would become invisible. At the surface of radius r_0 the term $(1 - 2GM/rc^2)$ in the Schwarzschild metric becomes zero; the curvature of space–time is so severe that we can only hope to give a consistent account of conditions using GR. For light travelling radially in a region described by the Schwarzschild metric, eqn (4.10), with $ds^2 = d\Omega^2 = 0$, becomes

$$0 = c^2\, dt^2(1 - r_0/r) - \frac{dr^2}{(1 - r_0/r)}.$$

Thus

$$c\, dt = \frac{dr}{1 - r_0/r}.$$

Then if a star shrinks to a radius less than r_0 the time taken for light to emerge from the spherical surface at r_0 becomes infinite, which confirms Michell's conjecture. However long an observer outside the radius r_0 waits he will never receive any light emitted from a source inside the radius r_0. For external observers this surface constitutes what is called an *event horizon*. The radius $r_0 \equiv 2GM/c^2$ is known as the *Schwarzschild radius* of the star. Therefore any star that shrinks within its Schwarzschild radius becomes invisible and is then described as a *black hole*.

The simplest (Schwarzschild) black holes have no angular momentum or charge. In Section 9.1 the space–time structure of Schwarzschild black holes is investigated, and the orbits for massive bodies circulating around them are calculated. The properties of black holes possessing some angular momentum are discussed briefly in Section 9.2. A paradoxical property of black holes is that they can radiate through a quantum field effect discovered by Hawking (1974). This Hawking radiation is discussed in Section 9.3. A star that is 1.4 times heavier than our Sun has a Schwarzschild radius of only 2 km, and if such a star were to shrink to this size it would attain a mean density of 10^{20} kg m^{-3}, far beyond the density of nuclei. It seems difficult to imagine how this could come about. However, Chandrasekhar (1931) deduced that the gravitational self-attraction of a sufficiently heavy star ($> 1.4\ M_\odot$) leads inevitably to its collapse to a point. Other physical processes can only check this collapse temporarily. This analysis is reviewed in Section 9.4.

The nature of black holes makes them difficult to detect; all we may ever have is indirect evidence for their existence. However, the available evidence that accretion onto a black hole powers the X-ray source Cygnus X-1 is compelling, and this case will be presented in Section 9.5.

9.1 The space–time structure

The event horizon at the radius $r_0 = 2GM/c^2$ is a fundamental feature of a black hole. Electromagnetic radiation originating within the horizon can never escape, so that space–time inside the horizon is effectively isolated from the rest of the Universe. In order to obtain a first idea of how space–time behaves across the horizon, consider an instrumented space probe falling along a radial path into a black hole. This black hole is taken to be a neutral non-rotating black hole (Schwarzschild black hole). Then space–time outside the parent body as it collapses to a point is described by the Schwarzschild metric: this region includes all space–time outside the horizon and all space–time down to the surface of the parent body inside the horizon. According to Chandrasekhar's analysis the parent star collapses completely to zero radius, and we shall generally take this to be the case. Equation (8.6) is the appropriate orbital equation. If the probe moves along a radius with $\theta = \pi/2$, $\varphi = 0$, then

$$\frac{(\mathrm{d}r/\mathrm{d}\tau)^2}{2} - \frac{GM}{r} = \frac{T}{m}.$$

For simplicity it can be imagined that the probe would have been be at rest if it had started at an infinite distance from the hole; then the kinetic energy parameter T vanishes and the equation reduces further to

$$c\,\mathrm{d}\tau = \pm\frac{r^{1/2}\,\mathrm{d}r}{r_0^{1/2}} \tag{9.1}$$

where, as usual, r_0 is $2GM/c^2$. Taking the negative sign for travel inward and then integrating gives

$$\tau = \tau_0 - \frac{r_0}{c}\left(\frac{2}{3}\right)\left(\frac{r}{r_0}\right)^{3/2} \tag{9.2}$$

where τ_0 is the proper time at which the probe reaches the centre ($r = 0$). This path is plotted in Fig. 9.1 as the solid curve. What is important to note is that the proper time τ as recorded by an on-board clock would change smoothly on crossing the horizon. Once across the horizon the probe soon reaches the centre of the hole; in the case of a hole of 10 $M_\odot$ this interval $\tau_0 - \tau$ is 10^{-4} s. In practice instruments could not survive such a journey; they would be torn apart by the increasing gravitational field gradients. Now consider how the same journey appears to a distant observer. The coordinate time t measured by this remote observer is related to the proper time through eqn (8.3):

$$\left(1 - \frac{r_0}{r}\right)\frac{\mathrm{d}t}{\mathrm{d}\tau} = K \text{ (a constant).}$$

Imposing the initial conditions, with the probe at rest and remote from the black hole, gives $K = 1$. Thus

$$\left(1 - \frac{r_0}{r}\right)\mathrm{d}t = \mathrm{d}\tau. \tag{9.3}$$

Equation (9.1) can now be used to replace $\mathrm{d}\tau$ in eqn (9.3); we have

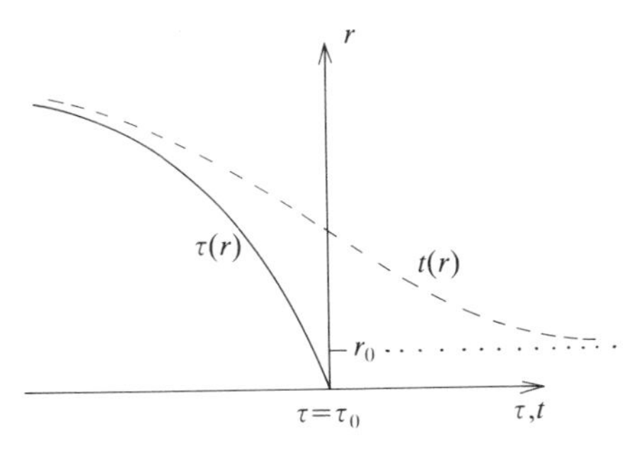

Fig. 9.1 The proper time τ and the coordinate time t plotted as a function of the radial coordinate r for a probe falling radially into a black hole.

$$c\left(1 - \frac{r_0}{r}\right) dt = -\frac{r^{1/2}\, dr}{r_0^{1/2}}$$

i.e.

$$c\, dt = -\frac{dr\, r^{3/2}}{(r - r_0) r_0^{1/2}}.$$

Integration for the inward journey yields

$$t = t_0 + \frac{r_0}{c}\left[-\frac{2}{3}\left(\frac{r}{r_0}\right)^{3/2} - 2\left(\frac{r}{r_0}\right)^{1/2} + \ln\left|\frac{(r/r_0)^{1/2} + 1}{(r/r_0)^{1/2} - 1}\right|\right]. \tag{9.4}$$

At large distances from the back hole this reduces to

$$t \approx t_0 - \frac{r_0}{c}\left(\frac{2}{3}\right)\left(\frac{r}{r_0}\right)^{3/2} - 2\left(\frac{r_0}{c}\right)\left(\frac{r}{r_0}\right)^{1/2}.$$

By choosing t_0 suitably it is possible to arrange that t and τ are equal at some large distance R. This choice is

$$t_0 \approx \tau_0 + 2\frac{r_0}{c}\left(\frac{R}{r_0}\right)^{1/2}.$$

From eqn (9.4) it is quite clear that as r tends toward r_0 then t tends to infinity. The world line described by eqn (9.4) is shown as the broken line in Fig. 9.1. The very different behaviour of coordinate time t and the proper time τ when the probe approaches and crosses the horizon illustrates vividly how space–time curvature makes it impossible to cover all space–time with one set of Cartesian coordinates.

If the probe emits signals at constant frequency (using, for example, an on-board quartz crystal oscillator), then, as it approaches the horizon, the photons travelling to the distant observer become less frequent and are increasingly red shifted. Both effects diminish the energy received so that eventually the signals are undetectable. By calculating the arrival time of photons reaching the distant observer it is possible to show that the energy in the signal fades away quickly. The photons in question follow radial paths for which the time elapsed between emission and detection is

$$t' = \int \frac{dr}{c(1 - r_0/r)}$$

$$= \frac{R - r}{c} + \frac{r_0}{c} \ln\left(\frac{R - r_0}{r - r_0}\right) \tag{9.5}$$

where the probe is at radius r and the observer at radius R. As far as the observer is concerned the arrival time T is measured relative to some fixed event, which can be the departure of the probe. This time interval is the sum of the times given by eqn (9.4) and (9.5). The expressions simplify a good deal once it is noted that the most significant contributions come from terms containing $\ln(r^{1/2} - r_0^{1/2})$ which are important when the probe passes across the region where $r \approx r_0$. Firstly from eqn (9.4)

$$t \approx -\left(\frac{r_0}{c}\right) \ln\left[\left(\frac{r}{r_0}\right)^{1/2} - 1\right] \approx -\frac{r_0}{c} \ln(r^{1/2} - r_0^{1/2})$$

$$\approx -\frac{r_0}{c} \ln(r - r_0),$$

where terms like $\ln(r^{1/2} + r_0^{1/2})$ and $\ln(r_0^{1/2})$ have been treated as negligible compared with $\ln(r^{1/2} - r_0^{1/2})$. Next, from eqn (9.5)

$$t' \approx -\left(\frac{r_0}{c}\right) \ln(r - r_0).$$

Thus the total elapsed time is

$$T = t + t' \approx -2\left(\frac{r_0}{c}\right) \ln(r - r_0). \tag{9.6}$$

Now the energy $L(T)$ received per unit time contains a factor $(1 - r_0/r)^{1/2}$ due to the increase in time interval between photons and another factor $(1 - r_0/r)^{1/2}$ due to their red shift. Thus

$$L(T) \propto \left(1 - \frac{r_0}{r}\right) = \frac{r - r_0}{r}$$

Using eqn (9.6) this becomes

$$L(T) \propto \exp\left(-\frac{cT}{2r_0}\right).$$

In the case of a star of mass $3\ M_\odot$ the time constant r_0/c is of the order 10^{-4} s, which means that when the signals from the probe begin to fade they do so quickly. This analysis applies equally well to a star in the act of collapsing through its horizon: it grows redder and fades on a time-scale of r_0/c.

The critical difference between space–time inside and outside the horizon lies in the sign reversal of the metric coefficients $g_{00} = (1 - r_0/r)$ and $g_{11} = -(1 - r_0/r)^{-1}$ at the surface $r = r_0$. In tabular form we have

	$r > r_0$	$r < r_0$
g_{00}	$+$	$-$
g_{11}	$-$	$+$.

Therefore if a small change in t is made at constant radius inside the horizon $(r < r_0)$,

$$\frac{ds^2}{c^2} = d\tau^2 = g_{00}\ dt^2 < 0.$$

This is opposite in sign to the effect of a similar small change in t outside the horizon ($r > r_0$), namely

$$\frac{\mathrm{d}s^2}{c^2} = \mathrm{d}\tau^2 = g_{00}\,\mathrm{d}t^2 > 0.$$

Inside the horizon a separation in coordinate time has become *space-like* ($\mathrm{d}s^2 < 0$). Similar considerations show that inside the horizon a separation in radial coordinate only has become *time-like* ($\mathrm{d}s^2 > 0$). The curvature is so intense that if we insist in using coordinates appropriate to distant flat space–time, then we find that space and time inside the horizon interchange the properties normally associated with them. This has many significant consequences. One is that if the coordinate r is held fixed for any length of time the interval $\mathrm{d}s^2$ is negative and *space-like*. This would require a massive body taking such a path to follow a space-like geodesic. Such a situation is logically impossible, and therefore static equilibrium inside the horizon must be impossible. Viewed from inside the horizon the external universe would appear equally strange. It is worth remembering here that space–time as seen from free fall would locally be a Minkowski space. However, the region over which this Minkowski frame matches actual space–time would be small because of the intense curvature and would become smaller as the centre of the black hole is approached.

The absence of any indication in the world line of the on-board clock that it is crossing the horizon was illustrated in Fig. 9.1 and shows that the horizon is not a *physical* singularity; rather, it is a *mathematical* singularity. What then is the import of the mathematical singularity with g_{11} diverging to infinity as r goes to r_0? The mathematical singularity arises because the set of coordinates imposed everywhere is best suited to regions of small curvature. A set of coordinates more appropriate to the locale of the black hole was invented by Eddington (1924) and rediscovered by Finkelstein (1958); the time coordinate t is replaced by $\tilde{t}$ such that

$$\tilde{t} = t + \frac{r_0}{c}\ln\left|\left(\frac{r}{r_0} - 1\right)\right|$$

and

$$\mathrm{d}\tilde{t} = \mathrm{d}t - \frac{\mathrm{d}r}{c(1 - r/r_0)}.$$

In terms of this new time coordinate the Schwarzschild metric becomes

$$\mathrm{d}s^2 = \left(1 - \frac{r_0}{r}\right)c^2\,\mathrm{d}\tilde{t}^2 - 2c\,\mathrm{d}r\,\mathrm{d}\tilde{t}\left(\frac{r_0}{r}\right) - \mathrm{d}r^2\left(1 + \frac{r_0}{r}\right) + r^2\,\mathrm{d}\Omega^2. \quad (9.7)$$

The metric coefficients of eqn (9.7) are no longer mathematically singular at the horizon. We shall show next that Eddington's coordinates (also called

Eddington–Finkelstein coordinates) provide the basis for a clearer physical understanding of space–time structure close to the black hole, although such coordinates would be a strange choice in a region remote from the black hole.

The radial path of a light ray in these new coordinates is

$$0 = \left(1 - \frac{r_0}{r}\right) c^2 \, \mathrm{d}\tilde{t}^2 - 2c \frac{r_0}{r} \, \mathrm{d}r \, \mathrm{d}\tilde{t} - \left(1 + \frac{r_0}{r}\right) \mathrm{d}r^2,$$

which has two solutions:

$$\frac{\mathrm{d}\tilde{t}}{\mathrm{d}r} = -\frac{1}{c} \qquad \text{and} \qquad \frac{\mathrm{d}\tilde{t}}{\mathrm{d}r} = \frac{1}{c} \frac{1 + r_0/r}{1 - r_0/r}. \tag{9.8}$$

These solutions describe the paths of ingoing and outgoing rays respectively, and in the absence of the black hole would become

$$\frac{\mathrm{d}\tilde{t}}{\mathrm{d}r} = -\frac{1}{c} \qquad \text{and} \qquad \frac{\mathrm{d}\tilde{t}}{\mathrm{d}r} = +\frac{1}{c}.$$

Light cones constructed according to eqn (9.8) are drawn in Fig. 9.2 at various points on the trajectory of a source falling into a black hole. The maximum inward component of the velocity of light is c throughout, while the maximum outward component varies with the radial distance. When the source is far from the horizon $\mathrm{d}\tilde{t}/\mathrm{d}r$ is $+1/c$ for light directed outwards. Then as the source

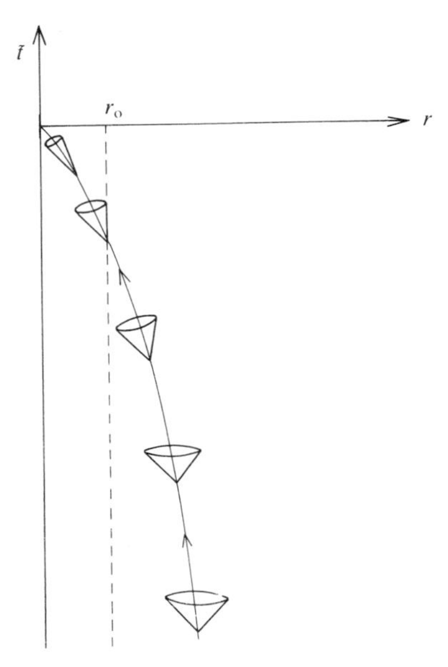

Fig. 9.2 The light cones of a probe falling into a black hole. $\tilde{t}$ is the Eddington coordinate time.

approaches the horizon $\mathrm{d}\tilde{t}/\mathrm{d}r$ increases until, at the horizon, it points along the time axis; the maximum outward component of velocity of light therefore falls to zero at the horizon. When the source goes inside the horizon $\mathrm{d}\tilde{t}/\mathrm{d}r$ becomes negative for 'outgoing' as well as ingoing rays so that the whole light cone is tilted inward. Although the velocity of light remains equal to c the future lies inward. This also means that once the horizon is crossed the source will inevitably head toward the centre of the black hole. However powerful the rocket power available may be, the probe's velocity vector must lie *inside* the light cone, and this seals its fate.

There exists an equally valid alternative choice for Eddington coordinates with

$$\tilde{t} = t - \frac{r_0}{c}\ln\left|\left(\frac{r}{r_0} - 1\right)\right|.$$

With this second choice for $\tilde{t}$ eqn (9.8) is replaced by solutions

$$\frac{\mathrm{d}\tilde{t}}{\mathrm{d}r} = -\frac{1}{c}\left(1 + \frac{r_0}{r}\right)\Big/\left(1 - \frac{r_0}{r}\right) \qquad \text{and} \qquad \frac{\mathrm{d}\tilde{t}}{\mathrm{d}r} = \frac{1}{c}. \tag{9.8$'$}$$

Now the light cone of a source within the horizon tilts so that it points *outward*. The physical situation is that of a 'white' hole ejecting material from the singularity at $r = 0$ into space–time. Gravitational collapse may create a black hole, but there is no physical process to generate its converse, the 'white' hole, although mathematically the situations are on an equal footing. The alternative coordinate choices show that space–time described by the Schwarzschild metric has many subtle properties. These were discussed further in a short seminal article by Kruskal (1960) and can be studied in the book by Misner, Thorne, and Wheeler (1972).

Conditions at the centre of the black hole ($r = 0$) cannot be described in any sensible manner as yet. The theory of general relativity predicts an infinite curvature, but whether such a physical singularity can really occur is not known. It is possible that quantum effects inhibit the formation of physical singularities (see Chapter 12).

The analysis of stable orbits developed in Chapter 8 can also be applied when the parent body is a Schwarzschild black hole. Combining eqns (8.4) and (8.5) to eliminate $\mathrm{d}\varphi/\mathrm{d}\tau$ gives

$$\frac{E^2}{Zc^2} - \frac{m^2(\mathrm{d}r/\mathrm{d}\tau)^2}{Z} - \frac{m^2J^2}{r^2} = m^2c^2.$$

where E is a constant of motion equivalent to the classical energy, and the angular momentum J per unit mass is another constant of motion. Rearrangement of the above equation gives

$$\frac{E^2}{m^2c^4} = \frac{(\mathrm{d}r/\mathrm{d}\tau)^2}{c^2} + Z\left(1 + \frac{J^2}{c^2r^2}\right) \tag{9.9}$$

which is the one-dimensional energy equation for radial motion. In eqn (9.9) the final term is an equivalent potential energy squared:

$$V^2(r) = Z\left(1 + \frac{J^2}{c^2r^2}\right). \tag{9.10}$$

This function is drawn in Fig. 9.3. At energies between E_1 and E_2 bound elliptic orbits are possible. For example, at an energy E, the orbit has a semi-major axis of length r_B and a semi-minor axis of length r_A. When the energy is exactly E_1 or E_2 the orbits are circular of radii r_1 and r_2 respectively, but only the orbit at radius r_1 is stable. r_1 and r_2 are the radii at which

$$\mathrm{d}V^2/\mathrm{d}r = 0$$

i.e.

$$r = r_J\left[1 \pm \left(1 - \frac{3r_0}{r_J}\right)^{1/2}\right], \tag{9.11}$$

where $r_J = J^2/c^2r_0$. r_2 is the smaller root and r_1 is the larger root. The radius of the *smallest* stable circular orbit is found by taking the factor under the square root sign to be zero:

$$1 - 3r_0/r_J = 0, \qquad \text{i.e. } J^2 = 3r_0^2c^2. \tag{9.12}$$

Then

$$r = r_J = 3r_0. \tag{9.13}$$

The energy of this smallest stable circular orbit is given by eqn (9.9) as

$$E^2(\text{min}) = \frac{8m^2c^4}{9}, \qquad \text{i.e. } E(\text{min}) = \left(\frac{8}{9}\right)^{1/2} mc^2. \tag{9.14}$$

Therefore the binding energy amounts to a fraction $1 - (8/9)^{1/2}$ or 5.72 per cent of the rest mass energy. This quantity is extremely important. It is the

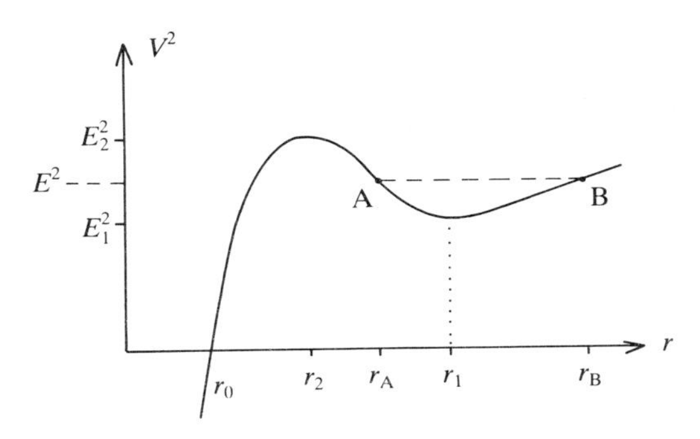

Fig. 9.3 The equivalent radial potential seen by a massive body in orbit around a Schwarzschild black hole. The labels are discussed in the text.

energy release when material accreted around a black hole spirals into the stable orbit with lowest energy.

By comparison the maximum energy release in thermonuclear fusion when hydrogen burns to ^{56}Fe is only 0.9 per cent of the rest mass energy. Gravitational energy release is therefore potentially the most important energy source in stellar processes. When a body orbits a rotating black hole the gravitational energy release can go much higher, reaching 42 per cent of the rest mass energy in favourable cases.

9.2 Rotating black holes

Anyone outside a black hole cannot follow what happens to material once it crosses the horizon. This raises the question of whether it is possible to make any distinction between black holes, beyond differences in mass. Hawking (1972) and others have proved rigorously that a full description of any black hole requires just three parameters; these are the total mass M, the total charge Q, and the total angular momentum or spin J of the black hole. In a neat phrase black holes are said to 'have no hair', which means that they have no detailed features.

As far as we know, matter on the large scale, and in particular stellar matter, appears to be nearly neutral. Therefore only *neutral rotating* black holes are considered here. The space–time around such a star is described by the Kerr metric. There are two important interfaces around a Kerr black hole. The first is the event horizon which is smaller than in the case of a non-rotating black hole; the radius is

$$r_- = \frac{r_0}{2} + \left[\left(\frac{r_0}{2}\right)^2 - a^2\right]^{1/2} \tag{9.15}$$

where $a = J/Mc$. The larger the angular momentum J of the star, the smaller the radius of the horizon becomes. In the limit that J/Mc exceeds $r_0/2$ the horizon would vanish and it would be possible to see the physical singularity at the centre of the black hole. What precisely would happen in the case of such a 'naked' singularity is not clear. Penrose has suggested that there is some physical principle ('cosmic censorship') which guarantees that all singularities lie safely within horizons. The second important surface defines a region outside the event horizon from which escape is possible but within which static equilibrium cannot be maintained. Any material in this intermediate region, which is known as the ergosphere, will rotate in the same sense as the black hole. The outer surface of the ergosphere is in fact a spheroid of revolution with a radial coordinate r_+ which is dependent on the polar angle θ with respect to the axis of rotation:

$$r_+(\theta) = \frac{r_0}{2} + \left[\left(\frac{r_0}{2}\right)^2 - a^2\cos^2\theta\right]^{1/2}. \tag{9.16}$$

At a radius $r_+(\theta)$ it is just possible to remain at rest. This surface touches the event horizon at the poles ($\theta = 0, \pi$) which is of course exactly where the effect of rotation vanishes. Penrose (1969) suggested a technique to extract energy from a rotating black hole. In one form this involves scattering radiation from the ergosphere; some energy from the rotation of the black hole is then transferred to the scattered radiation.

The arguments given above when taken together seem to lead to a contradiction. How is it, one may ask, that the electrostatic field due to the charge on the black hole penetrates through the horizon? After all it is well known that every type of electromagnetic effect is carried by photons, and we have seen that photons cannot escape from within a black hole. The answer to this question takes the discussion deep into the realm of quantum field theory. Usually we picture photons as transversely polarized; however, photons can equally well be polarized in the time direction or longitudinally along their direction of motion. When studying the propagation of electromagnetic waves (radiation) the effects of the 'longitudinal' and 'time' polarized photons cancel so that they can be ignored. However, they contribute to the *electrostatic* field and are not restrained by the horizon. Hence the charge within a black hole can exert an electrostatic field outside it. A similar view needs to be taken of the external gravitational field of the black hole if this is carried by massless field particles (gravitons) analogous to photons.

9.3 Hawking radiation

In 1974 Hawking made the astonishing discovery the black holes, previously regarded simply as absorbers of radiation, do in fact themselves radiate. This Hawking radiation has its origin in a quantum effect hitherto neglected in the previous discussion of black holes, namely that the vacuum is in a state of constant activity because of the creation and subsequent annihilation of particle–anti-particle pairs. For example a pair of photons can be created close to the black hole with four momenta (pc, $-\boldsymbol{p}$) and ($-pc$, $\boldsymbol{p}$). The net four-momentum is zero, but the negative energy photon violates the requirement that real photons have positive energy:

$$\text{energy} = +pc.$$

According to the Heisenberg uncertainty principle this 'virtual' photon can only exist for a time

$$\Delta t \sim \hbar/pc.$$

For some directions of emission the negative energy photon will cross the horizon. Once across the horizon and inside the black hole the space-like components of four-vectors become time-like, and vice versa. Thus the photon's negative energy converts to an acceptable spatial momentum and its momentum converts to an acceptable *positive* energy. Its lifetime is no longer restricted and so it travels quite freely within the black hole. The positive energy partner, which is a normal photon, travels freely outward, and such outgoing photons make up the Hawking radiation. A simple estimate of the temperature of this radiation can be obtained as follows, again using the uncertainty principle. The position of a photon emitted from the surface of the black hole is uncertain to $\sim r_0$. Accordingly the uncertainty in the photon momentum is

$$\Delta p \sim \hbar/r_0.$$

This momentum can be expressed in terms of a thermal energy kT at temperature T where k is the Boltzmann constant:

$$\Delta p \approx kT/c.$$

Combining the last two equations gives

$$\frac{kT}{c} \approx \frac{\hbar}{r_0} = \frac{\hbar}{2GM/c^2}.$$

Thus

$$T \approx \frac{\hbar c^3}{2kGM}.$$

This result can be compared with the exact expression obtained by Hawking. He has shown that a black hole of mass M radiates like a *black body* at a temperature

$$T = \frac{\hbar c^3}{8\pi kGM} \tag{9.17}$$

which only differs by a factor of 4π from the temperature deduced above. Putting in numbers we have

$$T = 6 \times 10^{-8}(M_\odot/\mathrm{M})\ \mathrm{K}.$$

The rate at which a black hole loses energy by Hawking radiation is

$$\frac{\mathrm{d}(Mc^2)}{\mathrm{d}t} = \sigma T^4(\text{surface area}) \propto M^{-3},$$

where σ is the Stefan–Boltzmann constant. Small black holes therefore have higher temperatures and radiate their energy more rapidly than larger black holes. The lifetime of a black hole is approximately given by

$$\tau \approx \left(\frac{M}{10^{11} kg}\right)^3 \times 10^{10} \text{ years.}$$

Thus any black hole of one solar mass would not have had time to evaporate since the origin of the Universe; however, very small black holes could have formed early in the life of the Universe and subsequently evaporated.

9.4 The formation of black holes

Two processes seem likely to lead to the formation of black holes. First, a star with a mass of roughly 20–60 $M_\odot$ will end its cycle of thermonuclear burning with an iron core which is massive enough to collapse under its own gravitational self-attraction. Second, the stars at the centre of galaxies may coalesce to form massive stars whose end product would be a super-massive black hole of mass 10^6–10^9 $M_\odot$. Experimental evidence on both processes is patchy, but there is at least one impeccable candidate for a black hole formed by the first mechanism, namely Cygnus X-1. Only the first mechanism will be discussed in detail.

A few comments will be made first about the life cycle of a star of mass roughly 20 $M_\odot$. Stars are initially formed from gas, mostly hydrogen, and contract under their own gravitational attraction. Eventually the internal pressure and temperature rise sufficiently for thermonuclear fusion to begin. The energy thus released sustains the star against collapse. Hydrogen is converted to helium, but once the core is completely converted further contraction takes place. This again raises the temperature and pressure, until the thermonuclear burning of helium to carbon is ignited. Subsequent stages of burning finally terminate with the star having an iron core. Iron nuclei have the largest binding energy per nucleon; hence neither fission nor fusion can release further energy. Burning at earlier stages continues in shells around the core, like the layers of an onion. By then several million years have elapsed since the birth of the star and its iron core contains several solar masses and has a density of about 10^{11} kg m^{-3} and a temperature of about 10^9 K. Further collapse is inhibited by the electron degeneracy pressure: this pressure is the direct consequence of the Pauli principle requirement that two electrons cannot occupy the same quantum state. However, Chandrasekhar (1931) showed that the degeneracy pressure is inadequate to resist collapse if the core mass exceeds 1.4 $M_\odot$. The argument can be understood as follows.

Suppose that a stellar core of mass M and radius R contains N electrons; their average spacing is thus

$$\Delta x \approx R/N^{1/3}.$$

The uncertainty in the individual electron momentum must be at least

$$p \approx \hbar/\Delta x \approx \hbar N^{1/3}/R.$$

Therefore if nearby electrons are to be in distinct quantum states their momenta must differ by at least p. This momentum is a measure of the average electron momentum if the gas of electrons is in its lowest energy state, i.e. it is degenerate. When the electrons are moving relativistically their individual energies are

$$E_\mathrm{R} \approx pc \approx \frac{\hbar N^{1/3} c}{R}, \tag{9.18}$$

while if their motion is non-relativistic their individual energies are

$$E_{\mathrm{NR}} \approx \frac{p^2}{2m_e} \approx \frac{\hbar^2 N^{2/3}}{2m_e R^2}. \tag{9.19}$$

The gravitational energy of the core is due almost entirely to nucleons of mass m_n. Therefore, assuming one nucleon per electron, the gravitational energy per electron is

$$E_G \approx -\frac{GNm_n^2}{R}. \tag{9.20}$$

It is now easy to determine the degeneracy pressure due to electrons. The first law of thermodynamics gives for adiabatic contraction

$$P = \frac{\mathrm{d}(NE)}{\mathrm{d}V} \propto \frac{N(\mathrm{d}E/\mathrm{d}R)}{R^2}.$$

Calculating $\mathrm{d}E/\mathrm{d}R$ from eqn (9.18) we see that, for relativistic electrons, the degeneracy pressure acting outwards is

$$P_R \propto R^{-4} \propto \rho^{4/3} \tag{9.21}$$

where ρ is the density of the core. However, using eqn (9.19) for non-relativistic electrons the degeneracy pressure acting outwards is

$$P_{\mathrm{NR}} \propto R^{-5} \propto \rho^{5/3} \tag{9.22}$$

Finally, using eqn (9.20) to calculate $\mathrm{d}E/\mathrm{d}R$ we find that the gravitational pressure inwards is

$$P_G \propto R^{-4} \propto \rho^{4/3}. \tag{9.23}$$

Equations (9.21), (9.22), and (9.23) allow us to constrast two regimes. In a core with low density the distinguishable electron quantum states are closely spaced in a momentum so that the degenerate electron gas is non-relativistic. Therefore eqn (9.22) holds, and because P_{NR} increases more rapidly with density than P_G, it is evident that at large enough density the pressures would reach equilibrium; the star would then stop shrinking. However, if the core mass is made larger, a condition is reached where the density and temperature are so large that the degenerate electron gas becomes relativistic. The electron gas is now described by eqn (9.21) which is a 'softer' equation of state. Contraction gives similar changes in P_R and P_G because they now both vary as $\rho^{4/3}$. Consequently once equilibrium is lost the contraction goes on indefinitely. The equilibrium state has

$$P_G = P_R \qquad \text{i.e. } E_R = |E_G|.$$

Using equations (9.18) and (9.20) we obtain

$$\frac{\hbar N^{1/3} c}{R} = \frac{GNm_n^2}{R},$$

or

$$N^{2/3} = \frac{\hbar c}{Gm_n^2}.$$

Equilibrium against collapse can only be maintained if the number of electrons (and hence nucleons) is less than

$$N(\text{max}) = \left(\frac{\hbar c}{Gm_n^2}\right)^{3/2} = 2 \times 10^{57}.$$

The mass of a core at this limit of stability is

$$M\ (\text{Chandrasekhar}) = N(\text{max})m_n = 1.4\ M_\odot.$$

Heavier cores continue to collapse with consequent rises in pressure and temperature. Then, when the mean electron energy reaches about 1 MeV, it is possible for electrons to initiate *neutronization* reactions such as

$$e^- + {}^{56}\text{Fe} \rightarrow {}^{56}\text{Mn} + \nu$$

which requires a threshold energy of 3.7 MeV. Neutronization is swift, taking about 1 s, and is accompanied by further core collapse. When the pressure reaches about 10^{17} kg m^{-3}, the nucleons are mostly free neutrons. The closely packed neutrons form a degenerate gas and, being fermions, they can also exert a degeneracy pressure. A repetition of the analysis just given shows that stellar cores of mass less than M(Chandrasekhar) are stable against further collapse and form neutron stars. Heavier cores inexorably collapse to form black holes. The critical core size can be estimated by making the approximation that the fermions of mass m_f are just becoming relativistic: $E \approx m_f c^2$. Substituting this value for E_R in eqn (9.18) gives

$$R = \frac{\hbar}{m_f c} N(\text{max})^{1/3}.$$

Taking $m_f = m_e$ gives an estimate for the limiting size of an iron star (white dwarf) as 5000 km. Taking $m_f = m_n$ gives an estimate of 3 km for the limiting size of a neutron star. This is about a factor of 3 smaller than current best estimates of the neutron star radius.

Thus far the history of the core has been followed. The most spectacular external manifestation of the process occurs when the collapse is checked by the stiffening of the core due to nuclear repulsion. Then the imploding outer layers of the star strike the core and rebound under the shock. These layers are ejected as a type II supernova explosion, of which SN1987a was a recent example.

Much effort has gone into refining the calculation of the limiting mass for a stable core with calculations being made in the framework of GR. In addition the effects of changing the equation of state relating pressure p to density ρ

have been explored. Oppenheimer and Volkoff (1939) assumed a spherically symmetric metric and a stress-energy tensor for the fluid core of the form

$$T_{\mu\nu} \doteq \left(\rho + \frac{p}{c^2}\right) v_\mu v_\nu - g_{\mu\nu} p$$

where v_μ is the four-vector fluid velocity (see Section 1.2). In free fall with Cartesian coordinates $T_{\mu\nu}$ has no off-diagonal elements and hence the fluid has no viscosity or thermal conductivity; it is an ideal fluid. The solution of Einstein's equation in this case is the Oppenheimer–Volkoff equation:

$$-r^2\left(\frac{\mathrm{d}p}{\mathrm{d}r}\right) = \frac{G}{c^4}(p + \rho c^2)\frac{Mc^2 + 4\pi r^3 p}{Z} \tag{9.24}$$

where M is the total mass given by

$$M = \int_0^r 4\pi r^2 \rho(r)\,\mathrm{d}r.$$

In the Newtonian limit $Z \approx 1$, $p \ll \rho c^2$ and $Mc^2 > r^3 p$ so that eqn (9.24) becomes:

$$-\frac{\mathrm{d}p}{\mathrm{d}r} = \frac{MG\rho}{r^2}. \tag{9.25}$$

Equation (9.25) can be obtained by equating the net radial pressure on a shell of thickness $\mathrm{d}r$ to the gravitational attraction $GM\rho\,\mathrm{d}r/r^2$ per unit area of the shell. Note that pressure as well as density contribute to the gravitational term in eqn (9.24) because pressure has a dual role in GR. It opposes collapse, but, as a component of the stress-energy tensor, it also contributes to hastening the collapse. The effect is to reduce the maximum stable mass a little. When rotation is taken into account the maximum stable mass is found to increase. In this case Friedman, Ipser, and Parker (1984), using the stiffest plausible equation of state for a star, obtain a maximum stable mass of 3.5 $M_\odot$. A radius of 10–15 km is predicted for such stars.

It is highly relevant to ask what values have been measured for neutron star masses; do they fall in with expectation? Only seven measurements, all on binary pulsars, have been made. These mass measurements cluster around 1.5 $M_\odot$. with errors of about 1 $M_\odot$. None demonstrably exceeds the mass limit of 3.5 $M_\odot$ for neutron stars obtained by Friedman *et al.*

9.5 Cygnus X-1

The observation of black holes must rely on indirect methods and hence an isolated black hole offers no prospects at all. A proportion of the X-ray pulsars, which are near relations of the black holes, are members of binaries, and so it is reasonable to expect that some stellar black holes should be bound to a visible star. Then the likelihood for detection is quite promising, as the case of Cygnus X-1 illustrates.

Cygnus X-1 emits X-rays and radio waves strongly with no signs of the regular time structure that marks out pulsars. An optical star HDE 226868 is located within the 1 arcsec error box defined by the radio measurements of the position of Cygnus X-1. This star has a spectrum that identifies it as a star of the type called blue supergiants, whose masses range from 20 to 100 $M_{\odot}$. It has been deduced from its apparent brightness that HDE 226868 lies about 2.5 kpc from the Earth. However, blue supergiants are too cold to be the source of the X-rays. Finally the wavelengths of the lines in the spectrum of HDE 226868 change with time following a regular cycle that repeats every 5.6 days.

An economical explanation of the features just presented is that HDE 226868 and Cygnus X-1 form a binary pair with an orbital period of 5.6 days: the time-varying spectral shift of the optical partner is then simply the Doppler shift produced by its orbital motion. Material from the blue supergiant can be transferred as shown in Fig. 9.4 to form an accretion disc around the black hole, a process that also occurs in binaries containing neutron stars. The accretion disc is drawn into the compact star by gravitational attraction with a large energy release (see Section 9.1). This process easily heats the accretion disc to a temperature at which it can emit X-rays.

The motion of HDE 226868 has been analysed by techniques similar to those already described in Section 8.6 for the pulsar 1913 + 16, this time using the Doppler spectral shift of HDE 226868. When this Doppler shift is converted to relative velocity it turns out that the velocity varies sinusoidally with time. As noted in Section 8.6 this shows that the orbits are *circular*. Suppose that the orbital radii are r_x and r_G for Cygnus X-1 and the blue supergiant respectively, and that the normal to the orbital plane is tilted at an angle i with respect to our line of sight. Then the maximum velocity of HDE 226868 toward us is

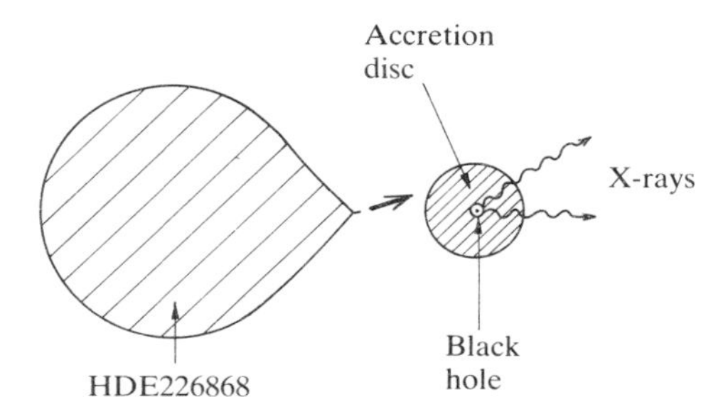

Fig. 9.4 A section taken in the orbital plane of the X-ray binary Cygnus X-1. The arrow indicates the flow of material from the blue supergiant HDE 226868 onto the accretion disc surrounding the black hole.

$$v_G = \frac{2\pi r_G}{\tau} \sin i \tag{9.26}$$

where τ is the orbital period. Measurements by Ninkov, Walker, and Yang (1987) give

$$\tau = 5.59964 \pm 0.00001 \text{ days} \qquad \text{and} \qquad v_G = 94.3 \pm 5 \text{ km s}^{-1}.$$

Using eqns (8.18) and (9.26) we can now determine of the mass function:

$$\frac{(m_x \sin i)^3}{(m_x + m_G)^2} = \frac{(r_G \sin i)^3 (2\pi/\tau)^2}{G}$$

$$= \frac{{v_G}^3 \tau}{2\pi G}$$

$$= 0.484\ M_\odot.$$

A lower limit on the mass of Cygnus X-1 can now be obtained by setting m_G to 20 $M_\odot$ and i to 90°; then

$$m_x \geqslant 10\ M_\odot,$$

which is already above the upper limit (3.5 $M_\odot$) for the mass of a neutron star. In fact i has to be less than 90° for the simple reason that the supergiant does not eclipse the X-ray emission from Cygnus X-1 at any part of the orbit; if i were 90° the orbits would be seen edge-on and eclipsing would be unavoidable. Making plausible assumptions about the size of HDE 226868, a limit can be imposed on i, and hence an improved lower limit obtained for m_x. This limit is 15 $M_\odot$, which makes it highly probable that Cygnus X-1 is indeed a black hole. Two other good candidates for black holes are also X-ray binaries: LMC X-3 and A0620-00. McClintock (1988) quotes the most probable mass values for these three cases as 16 $M_\odot$, 9 $M_\odot$ and 9 $M_\odot$ respectively. It is worth nothing that the X-ray emission from Cygnus X-1 is irregular and flickers on a millisecond time-scale. This is at least consistent with the source being less than one milli-light-second (i.e. 300 km) across.

To have accumulated only a few examples of candidate black holes might sound rather a thin result. It should be remembered, however, that among the few hundred pulsars detected only a few are binary pulsars; binary black holes may therefore form a comparably small proportion of all black holes. Added to this the partner of the black hole must have swollen to the giant stage in order that material can be transferred rapidly enough to give an accretion disc that produces detectable numbers of X-rays. Finally there is the complication that X-ray emissions can only be detected from above the Earth's atmosphere. Small wonder then that there are only a few candidates!

10
Gravitational radiation

Einstein showed that the existence of gravitational radiation is a natural consequence of the general theory of relativity. He considered small disturbances from a flat space–time, which is appropriate for the waves likely to reach Earth. In this limit the Einstein equation reduces to a linear wave equation. This has plane wave solutions which are transverse waves travelling with velocity c, properties that electromagnetic waves also possess. During the passage of gravitational waves it is the structure of space–time itself which oscillates. Putting this more precisely, the proper time taken by light to pass to and fro between two fixed points in spaces oscillates. There is no effect at a single point, only a change in the *separation* of points in space. In contrast, the electromagnetic E and B fields measured at a point oscillate when an electromagnetic wave passes. Electromagnetic radiation from a source small compared with the wavelength is predominantly dipole in character with higher multipoles becoming progressively weaker. For gravitational radiation the same effect is present, but because dipole radiation is forbidden by conservation laws, the dominant gravitational waves are quadrupole.

The constant $c^4/8\pi G$ appearing in Einstein's equation can be interpreted as the force per unit area required to give space–time unit curvature, that is 10^{43} N m^{-2} for a curvature of 1 m^{-2}. Space–time is therefore an extremely stiff medium, and by the same token small-amplitude waves carry large energies. The most energetic waves likely to be observed on Earth are those due to stellar collapse in our Galaxy. Gravitational waves emanating from such events are predicted to produce stresses in space–time of amplitude about 10^{-18}. This represents a displacement of under one nuclear diameter in a length of 1 m. The measurement of such a tiny strain is a challenge to which experimentalists have vigorously responded. Two designs for detectors of gravitational radiation are discussed below. Weber, who initiated experimental searches in 1960, used a freely suspended massive bar whose suspension is isolated from external mechanical vibrations. A gravitational wave at the natural frequency for longitudinal oscillations of the bar would set it ringing like a tuning fork. Any ringing would be detected by transducers (e.g. piezoelectric crystals) which are mechanically connected to the bar. Such bar detectors are called resonant detectors. The other class of devices are designed to detect the relative movement of freely suspended massive blocks. If their suspensions are isolated from external disturbance the blocks will respond faithfully to the horizontal component of any incident gravitational wave. A preferred method for detecting the relative movement is to build a Michelson interferometer with the mirrors mounted on these freely suspended blocks. Quadrupole radiation incident with a favourable polarization should cause the two arm lengths to expand and contract in anti-phase. The difference between the two optical path lengths would oscillate and produce an oscillation of the interference fringe pattern, and this is what the experimentalists aim to detect. To date there has been no direct detection of gravitational waves by either method. The sole observation is indirect; the orbital period of the binary pulsar 1913 + 16 is measured to be slowing down at a rate which agrees well with that expected if the system is emitting gravitational radiation.

Properties of gravitational waves are treated in Section 10.1, and the effects of gravitational waves in Section 10.2. PSR 1913 + 16 is discussed in Section 10.3. In Section 10.4 estimates are given for the intensity of gravitational radiation on Earth from various sources; finally in Section 10.5 current detector designs, which appear capable of detecting waves from supernovae in our Galaxy, are described.

10.1 Properties of gravitational radiation

The basic equation to describe empty space–time is eqn (7.19) with the right-hand side set to zero:

$$G_{\beta\delta} = R_{\beta\delta} = 0 \tag{10.1}$$

When the space–time curvature is small, a linear approximation can be made to the metric

$$g_{\beta\delta} = \eta_{\beta\delta} + h_{\beta\delta} \tag{10.2}$$

where $\eta_{\alpha\beta}$ is the Minkowski metric and all the components $h_{\beta\delta}$ are very much less than unity. $g_{\beta\delta}$ and $\eta_{\beta\delta}$ are symmetric tensors, and hence $h_{\beta\delta}$ is also a symmetric tensor. What we shall do is solve

$$G^{(1)}_{\beta\delta} = R^{(1)}_{\beta\delta} = 0 \tag{10.1a}$$

where only terms linear in $h_{\beta\delta}$ are retained. The metric connections can be obtained from eqn (6.5), while noting that the differentials of $\eta_{\mu\nu}$ are zero:

$$2\Gamma^{\alpha}_{\beta\delta} = 2g^{\alpha\nu}\Gamma_{\nu\beta\delta} = g^{\alpha\nu}(h_{\beta\nu,\delta} - h_{\delta\beta,\nu} + h_{\nu\delta,\beta}).$$

To first order in $h_{\alpha\beta}$, we obtain with the help of eqn (7.3)

$$R_{\beta\delta} = R^{\alpha}_{\beta\alpha\delta} = \Gamma^{\alpha}_{\beta\delta,\alpha} - \Gamma^{\alpha}_{\beta\alpha,\delta},$$

excluding the products in Γ because they are of order h^2. Then to the same approximation

$$R_{\beta\delta} = \tfrac{1}{2}g^{\alpha\nu}(h_{\nu\delta,\beta\alpha} - h_{\delta\beta,\nu\alpha} + h_{\alpha\beta,\nu\delta} - h_{\alpha\nu,\beta\delta}).$$

Substituting this value into eqn (10.1a) yields

$$h^{\alpha}{}_{\delta,\beta\alpha} - h_{\delta\beta,}{}^{\alpha}{}_{\alpha} + h_{\alpha\beta,}{}^{\alpha}{}_{\delta} - h^{\alpha}{}_{\alpha,\beta\delta} = 0 \tag{10.3}$$

where we have used the fact that $g^{\alpha\nu}$ raises the subscripts of h. From any solution of this equation we can generate equivalent solutions by changing coordinates; they are equivalent because the measurable tidal effects are determined by the difference in h at two places and remain the same whatever the choice of coordinates. One simple solution which is easy to interpret is obtained by requiring that each term in the wave equation is separately zero. Our simplification is to have

$$h^{\alpha}{}_{\alpha} = 0 \tag{10.3a}$$

$$h^{\alpha}{}_{\delta,\alpha} = h_{\alpha\beta,}{}^{\alpha} = 0 \tag{10.3b}$$

$$h_{\beta\delta,}{}^{\alpha}{}_{\alpha} = 0. \tag{10.3c}$$

Equation (10.3a) requires the trace to vanish and (10.3b) requires the divergence to vanish, so that any solution of eqn (10.3c) is a solution of (10.3) provided that it is traceless and its divergence vanishes. One final requirement can be made on the coordinate choice

$$h_{\alpha 0} = 0. \tag{10.3d}$$

The physical significance of this choice will be apparent when a solution of (10.3c) is written down. The condition (10.3b) contains four constraints. Equation (10.3a) plus (10.3d) impose four additional constraints when allowance is made for the symmetry of $h_{\alpha\beta}$. In general $h_{\alpha\beta}$ has ten independent components so that when these eight constraints are imposed there remain just two independent components of $h_{\alpha\beta}$. These emerge below as the two polarization states of eqn (10.11). Equation (10.3c) is simply

$$\frac{\partial^2 h_{\beta\delta}}{\partial x_{\alpha}\partial x^{\alpha}} = 0, \tag{10.4}$$

i.e.

$$\frac{\delta^2 h_{\beta\delta}}{(\partial ct)^2} = \frac{\partial^2 h_{\beta\delta}}{\partial x^2} + \frac{\partial^2 h_{\beta\delta}}{\partial y^2} + \frac{\partial^2 h_{\beta\delta}}{\partial z^2}. \tag{10.4a}$$

Equation (10.4) is the familiar wave equation for waves with velocity c. A plane wave solution for a wave travelling along the direction given by the unit vector $\boldsymbol{n}$ in *space* is

$$h_{\beta\delta} = A_{\beta\delta}\exp(ikn_{\alpha}x^{\alpha}) \tag{10.5}$$

where k is the wave number. Substitution for $h_{\beta\delta}$ in eqn (10.4) gives

$$n^{\alpha}n_{\alpha} = 0$$

so that n^{α} is light-like, and hence $n^{\alpha} = (1, \boldsymbol{n})$. Specializing to a wave travelling along the z axis,

$$h_{\beta\delta} = A_{\beta\delta} \cos[k(ct - z)]$$
$$= A_{\beta\delta} \cos(\omega t - kz). \tag{10.6}$$

Here $\lambda = 2\pi/k$ is the wavelength and $\omega = kc$ is the angular frequency of the wave. These solutions must satisfy the subsidiary conditions of eqns (10.3a), (10.3b), and (10.3d). Now

$$h_{\delta\beta,}{}^{\delta} = \frac{\partial h_{\beta\delta}}{\partial x_\delta} = ikA_{\beta\delta}n^{\delta}.$$

Thus making use of eqn (10.3b) we have $A_{\beta\delta}n^{\delta} = 0$ which, for a wave travelling in the $z(3)$ direction, becomes

$$A_{\beta 0} + A_{\beta 3} = 0 \qquad \text{for all } \beta.$$

Equation (10.3d) then gives

$$A_{\beta 0} = 0 \qquad \text{for all } \beta.$$

Combining the last two results, we have

$$A_{\beta 3} = 0 \qquad \text{for all } \beta.$$

Finally we can use the fact that $h_{\alpha\beta}$ is symmetric, giving

$$A_{0\beta} = A_{\beta 0} = A_{\beta 3} = A_{3\beta} = 0. \tag{10.7}$$

Therefore the wave only has non-zero components along the $x(1)$ and $y(2)$ directions and is transverse to the direction of motion $z(3)$. The trace condition of eqn (10.3a) gives

$$A_{11} + A_{22} = 0. \tag{10.8}$$

Using the symmetry of $h_{\alpha\beta}$ again we also have

$$A_{12} = A_{21}. \tag{10.9}$$

It is now easy to write $A_{\alpha\beta}$ in full using eqns (10.7), (10.8), and (10.9):

$$A_{\beta\delta} = \begin{matrix} 0 & 0 & 0 & 0 \\ 0 & A_{11} & A_{12} & 0 \\ 0 & A_{12} & -A_{11} & 0 \\ 0 & 0 & 0 & 0. \end{matrix} \tag{10.10}$$

The choice of coordinates has made the wave amplitudes $A_{\beta\delta}$ both *transverse* and *traceless*; therefore the restriction on the coordinates is called the transverse–traceless gauge. Notice that eqn (10.3d) has the effect of requiring the wave to be orthogonal to the time axis as well as to the direction of propagation in space. The general solution of the form used in eqn (10.6) is made up of a linear combination of the two orthogonal states:

$$h_{\beta\delta} = h_+(e_+)_{\beta\delta} \cos(\omega t - kz) \tag{10.11a}$$

and

$$h_{\beta\delta} = h_{\times}(e_{\times})_{\beta\delta} \cos(\omega t - kz + \varphi) \quad (10.11b)$$

where φ is an arbitrary phase angle and

$$(e_{+})_{\beta\delta} = \begin{matrix} 0 & 0 & 0 & 0 \\ 0 & 1 & 0 & 0 \\ 0 & 0 & -1 & 0 \\ 0 & 0 & 0 & 0 \end{matrix} \quad (10.11c)$$

$$(e_{\times})_{\beta\delta} = \begin{matrix} 0 & 0 & 0 & 0 \\ 0 & 0 & 1 & 0 \\ 0 & 1 & 0 & 0 \\ 0 & 0 & 0 & 0 \end{matrix} \quad (10.11d)$$

and where $h_{+} = A_{11}$, $h_{\times} = A_{12}$. e_{+} and $e_{\times}$ are two polarization states of gravitational radiation. Their tensor form implies a more complicated polarization than the linear polarization met with in the case of light. At this point we need to investigate the motion of nearby test masses lying in the path of a gravitational wave in order to be able to construct a picture of the effects of these two polarization states.

10.2 The effects of gravitational waves

Consider what happens in the plane $z = 0$ to two nearby bodies located at $A(x = \xi, y = 0)$ and $B(x = y = 0)$. Their proper separation

$$\begin{aligned} ds = \xi' &\approx [|g_{11}(t, 0)|]^{1/2}\xi \\ &\approx \left[1 + \frac{h_{11}(t, 0)}{2}\right]\xi. \end{aligned} \quad (10.12)$$

Thus applying eqn (10.11a) the proper time interval between A and B undergoes a strain of amplitude

$$\varepsilon = \frac{\xi' - \xi}{\xi} = \frac{h_{11}(0, 0)}{2} = \frac{h_{+}}{2}. \quad (10.13)$$

Thus $h_{+}/2$ is the amplitude of the differential (tidal) change in lengths between nearby points along the x direction. Suppose now that B $(l/2, 0)$ is a large distance from A $(-l/2, 0)$. Then their proper separation wll be

$$\begin{aligned} l(t) &= \int dx[g_{11}(t - x/c, 0)]^{1/2} \\ &= \int dx[1 + \tfrac{1}{2}h_{11}(t - x/c, 0)]. \end{aligned}$$

Using eqn (10.11a) this gives

$$l(t) = l_0 + \frac{h_+}{2} \int \mathrm{d}x \cos(\omega t - kx)$$

where l_0 is the separation before the gravitational wave arrived. Then the change in path length

$$\Delta l(t) = \frac{h_+}{2k} \left[\sin\left(\omega t + \frac{kl_0}{2}\right) - \sin\left(\omega t - \frac{kl_0}{2}\right) \right]$$

$$= \frac{h_+}{k} \cos(\omega t) \sin\left(\frac{kl_0}{2}\right).$$

The amplitude of this displacement is

$$\Delta l = \frac{\lambda h_+}{2\pi} \sin\left(\frac{\pi l_0}{\lambda}\right). \tag{10.14}$$

Therefore the strain amplitude is

$$\varepsilon = \frac{\Delta l}{l_0} = \frac{\lambda h_+}{2\pi l_0} \sin\left(\frac{\pi l_0}{\lambda}\right)$$

which reproduces eqn (10.13) when $l_0 \to 0$. Along the y direction we obtain similar results; in comparison with (10.13) the amplitude of the tidal effect along the y direction is

$$\varepsilon = -h_+/2. \tag{10.15}$$

Therefore test masses placed close to the origin move in the plane perpendicular to the wave direction with

$$x = x_0\left(1 + \frac{h_+ \cos \omega t}{2}\right)$$

$$y = y_0\left(1 - \frac{h_+ \cos \omega t}{2}\right)$$

for the e_+ state of polarization. This motion is shown on the left of Fig. 10.1 for a set of test masses originally at rest in a circle in the xy plane: their displacements are followed over a complete cycle. There are two orthogonal symmetry axes and so the motion is quadrupole. The movement obtained with the orthogonal polarization state of eqn (10.11b) with $\varphi = 0$ is

$$x = x_0 + \frac{y_0 h_\times \cos \omega t}{2}$$

$$y = y_0 + \frac{x_0 h_\times \cos \omega t}{2},$$

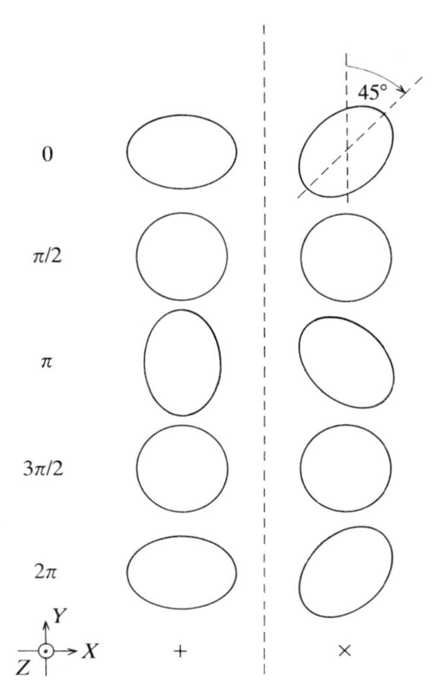

Fig. 10.1 The effect of gravitational waves on a circle of test masses followed over one cycle. The observer is looking towards the source. Two orthogonal states of quadrupole radiation are illustrated.

which can be manipulated to give

$$x + y = \left(1 + \frac{h_\times \cos \omega t}{2}\right)(x_0 + y_0)$$

and

$$x - y = \left(1 - \frac{h_\times \cos \omega t}{2}\right)(x_0 - y_0).$$

A complete cycle of movement is drawn on the right of Fig. 10.1. It is not possible to construct the e_+ pattern from the $e_\times$ pattern or vice versa; they are orthogonal polarization states. By analogy with electromagnetic waves the e_+ and $e_\times$ amplitudes can be added with phase differences of $-\pi/2\,(+\pi/2)$ to obtain right (left) circularly polarized amplitudes. These are

$$h_{\beta\delta} = h[(e_+)_{\beta\delta} \cos(\omega t - kz) \pm (e_\times)_{\beta\delta} \sin(\omega t - kz)].$$

Their effects are shown in Fig. 10.2 for the test mass arrangement described earlier. The arrows show the sense of rotation of the patterns. We can also draw diagrams complementary to Figs 10.1 and 10.2 which show tidal accel-

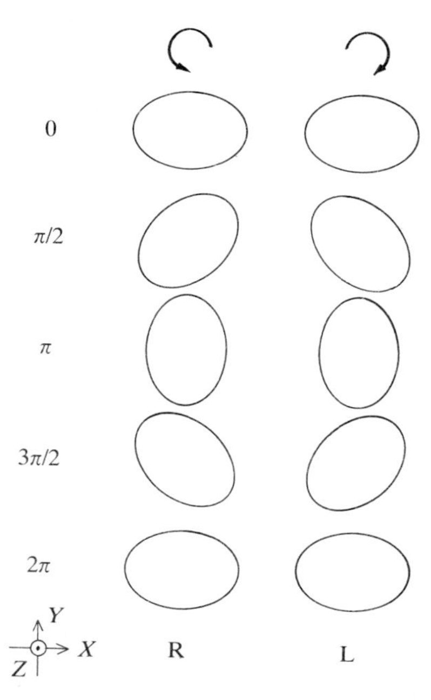

Fig. 10.2 The effect of gravitational waves on a circle of test masses followed over one cycle. The observer is facing the source. Right-handed and left-handed polarized quadrupole radiation patterns are illustrated.

erations. Just two examples are given in Fig. 10.3: for e_+ and $e_\times$ when the phase $\omega t - kz$ is zero. Not surprisingly these force patterns resemble the pattern of magnetic field lines across the aperture of a quadrupole magnet.

A revealing difference between gravitational and electromagnetic radiation is that dipole radiation is absent in the gravitational case. To understand why this should be we start by making a comparison between potentials: one is an electrostatic potential due to charges q^α at $\boldsymbol{r}^\alpha$, and the other is a gravitational potential due to masses m^α at $\boldsymbol{r}^\alpha$. For simplicity we take the nearly Newtonian case of slowly moving masses/charges. At a vector distance $\boldsymbol{R}$ from the centre of mass

$$4\pi\varepsilon_0 \varphi_{\text{es}}(\boldsymbol{R}) = \sum q^\alpha/|\boldsymbol{R} - \boldsymbol{r}^\alpha|$$

which can be expanded if $r \ll R$ as

$$4\pi\varepsilon_0 \varphi_{\text{es}} = \sum \frac{q^\alpha}{R} - \sum q^\alpha x_i^\alpha \left[\frac{\partial}{\partial X_i}\left(\frac{1}{R}\right)\right] + \cdots,$$

where $x_i^\alpha(X_i)$ are components of $\boldsymbol{r}^\alpha(\boldsymbol{R})$. Similarly, the gravitational potential is

$$\frac{\varphi}{G} = \sum \frac{m^\alpha}{R} - \sum m^\alpha x_i^\alpha \left[\frac{\partial}{\partial X_i}\left(\frac{1}{R}\right)\right] + \cdots.$$

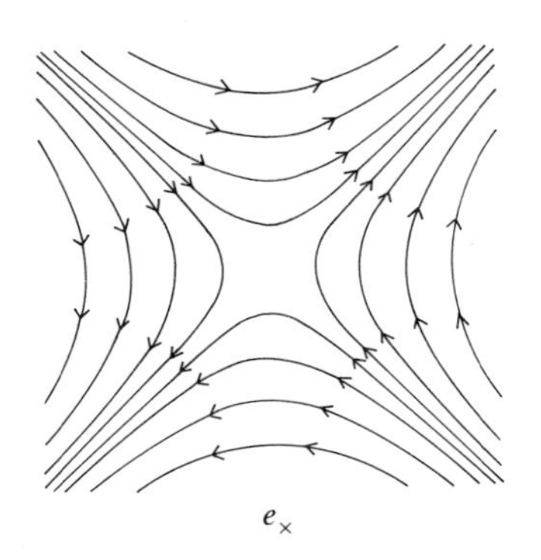

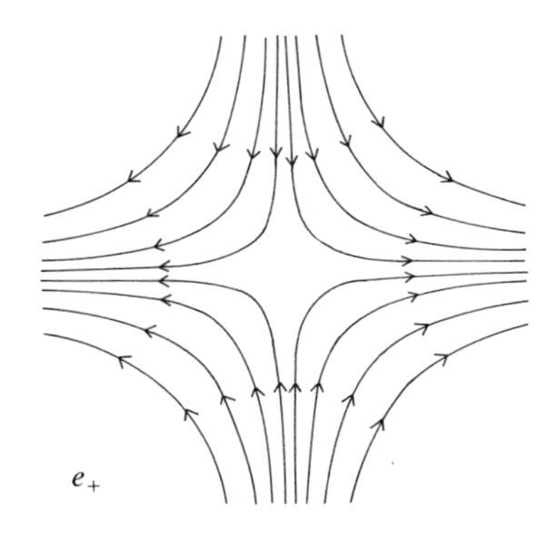

Fig. 10.3 The patterns of tidal acceleration for the two orthogonal states of polarization (e_+ and $e_\times$) when the phase angle is zero.

The radiation is proportional to $\mathrm{d}^2\varphi/\mathrm{d}t^2$ and the second term in the expansion is responsible for dipole radiation. In the gravitational case the dipole term contains a factor $\sum m^\alpha \mathrm{d}^2(x_i^\alpha)/\mathrm{d}t^2$ which equals the net force, and this vanishes for an isolated system. It is quite the opposite for a charge distribution which can easily have a non-zero oscillating electric dipole moment $\sum q^\alpha x_i^\alpha$. Similar considerations can be applied to 'magnetic' dipole radiation which is important when the charges are in rapid motion; it also vanishes identically in the gravitational case. This makes the next term in the potential, containing the quadrupole moment of the source $\sum m^\alpha x_i^\alpha x_j^\alpha$, the important one in generating gravitational waves.

Gravitational waves carry energy which is used to deform space–time. Therefore eqn (10.1a) needs to be modified in the presence of such waves, even though space–time is empty:

$$G_{\alpha\beta}^{(1)} = \left(\frac{8\pi G}{c^4}\right) t_{\alpha\beta}$$

where $t_{\alpha\beta}$ is the stress-energy tensor carried by the waves. Then

$$G_{\alpha\beta}^{(1)} - \frac{8\pi G}{c^4} t_{\alpha\beta} = 0.$$

This result is equivalent to the expansion of (10.1) to second order in $h_{\alpha\beta}$, namely

$$G_{\alpha\beta}^{(1)} + G_{\alpha\beta}^{(2)} = 0.$$

Thus to a good approximation we have

$$t_{\alpha\beta} = -\frac{c^4}{8\pi G} G_{\alpha\beta}^{(2)}. \tag{10.16}$$

This expression is evaluated in Appendix E. In the transverse traceless gauge

$$t_{00} = \frac{c^2}{16\pi G}\langle \dot{h}_+^2 + \dot{h}_\times^2 \rangle \tag{E.7}$$

and the energy flow per unit area per unit time is

$$F = \frac{c^3}{16\pi G}\langle \dot{h}_+^2 + \dot{h}_\times^2 \rangle \tag{E.8}$$

$$= \frac{c^3}{32\pi G}\langle \dot{h}_{ij}\dot{h}_{ij} \rangle. \tag{E.9}$$

It is misleading to assign energy to a point in space–time because only *relative* displacements are meaningful. Furthermore it is not even possible to specify whether the energy is in the peaks or valleys of the waves. For these reasons eqns (E.7), (E.8), and (E.9) only apply when averages are taken over several cycles and wavelengths. This averaging is indicated by the angular brackets.

Solutions of Einstein's equation will be sought for sources that are nearly Newtonian, which means firstly that within the source the curvature and strain are small and secondly that the velocity v of the material within the source is very much less than c. A linear approximation can be made to the metric in these circumstances. It turns out that the intensity of gravitational radiation predicted by this linearized theory generally only differs by small numerical factors from the results of more exact calculations (Davis *et al.* 1971). In Appendix F the contribution due to quadrupole motion of the source is evaluated. At a distance r from a source

$$h_{ij}(t) = \frac{2G}{rc^4}\ddot{I}_{ij}^{\mathrm{TT}}\left(t - \frac{r}{c}\right) \tag{F.4}$$

where I_{ij}^{TT} is the transverse–traceless part of the quadrupole moment of the source (see eqn (F.3)). Equation (F.4) is a *retarded* solution: gravitational disturbances propagate at a velocity c, and hence the amplitude at r at time t is determined by the source behaviour at an earlier time $t - r/c$. This argument will be omitted from hereon in order to simplify the presentation. The energy flow in the gravitational waves is obtained using eqn (E.9), giving

$$F = \frac{G}{8\pi r^2 c^5}\langle \dddot{I}_{ij}^{\mathrm{TT}}\dddot{I}_{ij}^{\mathrm{TT}} \rangle. \tag{F.5}$$

As before the angular brackets indicate the expectation value averaged over several cycles. The total energy flow through a sphere at a distance r from the source is the total energy output of the source, which is also called its *luminosity L*. In Appendix F it is shown that

$$L = \left(\frac{G}{5c^5}\right)\langle \dddot{I}_{ij}\dddot{I}_{ij} \rangle \tag{F.6}$$

where $I\!\!\!/_{ij}$ is the reduced quadrupole moment of the source

$$I\!\!\!/_{ij} = \int (x_i x_j - \delta_{ij} x_k^2/3)\rho \, \mathrm{d}V$$

with ρ being the density in a volume element dVat x_i. The range of integration is over the volume of the source. In the above equation δ_{ij} is the Kronecker delta, with a value $+1$ when $i = j$ and zero otherwise.

10.3 PSR 1913+16

The pulsar PSR 1913+16 and its compact companion form a binary pair whose orbital parameters are well known, thanks to the work of Taylor and his colleagues (see Section 8.6). It is therefore possible to calculate the quadrupole moment of this system and to infer its output of gravitational radiation. For orientation, consider two point masses M in circular orbits at a separation $2a$. Take the centre of mass as origin and the orbit to lie in the xOy plane with the masses along the x axis at time zero. Then the quadrupole moment has an xx component (see Question 10.1)

$$I_{xx} = 2Ma^2 \cos^2 \omega t = Ma^2(1 + \cos 2\omega t)$$

where ω is the angular frequency. Similarly

$$I_{yy} = Ma^2(1 - \cos 2\omega t).$$

Working in two dimensions (x, y) the *traceless* or reduced quadrupole moments are

$$I\!\!\!/_{ij} = \int (x_i x_j - \delta_{ij} x^2{}_k/2)\rho \, \mathrm{d}V$$

Thus

$$I\!\!\!/_{xx} = -I\!\!\!/_{yy} = Ma^2 \cos 2\omega t,$$

while

$$I\!\!\!/_{xy} = I\!\!\!/_{yx} = Ma^2 \sin 2\omega t.$$

Substituting these values into eqn (F.6) for the luminosity gives

$$L = \frac{G}{5c^5}(2\omega)^6 (Ma^2)^2 \langle 2 \sin^2 2\omega t + 2 \cos^2 2\omega t \rangle$$

$$= \frac{G}{5c^5}(128\omega^6 M^2 a^4).$$

In order to estimate the rate at which the orbit decays we need to compare this, the rate of energy loss, with the total energy of the binary pair

$$E = Mv^2 - GM^2/2a.$$

Using the radial equation of motion for either star

$$Mv^2/a = GM^2/4a^2$$

we obtain

$$v^2 = GM/4a,$$

whence

$$\omega^2 = v^2/a^2 = GM/4a^3.$$

Now, substituting for v in the expression for total energy and then rewriting a in terms of ω gives

$$E = -\frac{GM^2}{4a} = -\frac{GM^2}{4}\left(\frac{4\omega^2}{GM}\right)^{1/3}.$$

Thus

$$\frac{\mathrm{d}E}{E} = \frac{2}{3}\frac{\mathrm{d}\omega}{\omega} = -\frac{2}{3}\frac{\mathrm{d}\tau}{\tau}$$

where τ is the orbital period. Then the observable quantity, the orbital decay rate, is

$$\frac{(\mathrm{d}\tau/\mathrm{d}t)}{\tau} = \frac{3}{2}\frac{\mathrm{d}E/\mathrm{d}t}{E} = \frac{3}{2}\frac{L}{E} = -\frac{768}{5}\frac{\omega^6 a^5}{c^5}.$$

Substituting for ω gives finally

$$\frac{\mathrm{d}\tau/\mathrm{d}t}{\tau} = -\frac{12}{5}\frac{G^3M^3}{c^5a^4}. \tag{10.17}$$

Press and Thorne (1972) calculated the correction required for the case of an elliptical orbit of eccentricity e; the right-hand side of eqn (10.17) is then multiplied by a factor

$$\frac{1 + 73e^2/24 + 37e^4/96}{(1 - e^2)^{7/2}}.$$

Inserting the values of the orbital parameters measured by Taylor and his colleagues, given in Section 8.6, yields

$$\frac{\mathrm{d}\tau}{\mathrm{d}t} = -2.403(2) \times 10^{-12}.$$

The precision of measurement of the orbital period decay rivals the precision of this prediction. Taylor and his colleagues find

$$\frac{\mathrm{d}\tau}{\mathrm{d}t} = -2.40(9) \times 10^{-12},$$

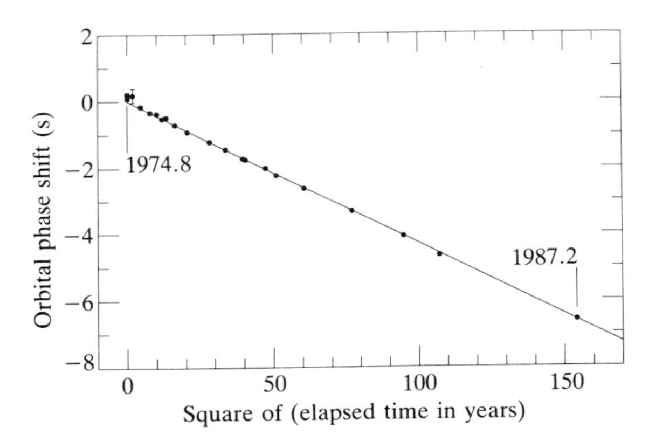

Fig. 10.4 The measured rate of lag of the orbital phase for the binary pulsar PSR 1913 + 16 as a function of time squared. The prediction of the GR is shown as a solid line. (J. H. Taylor and J. M. Weisberg, unpublished data, by courtesy of Professor Taylor.)

in convincing agreement with the prediction from GR. Figure 10.4 shows the decay of the orbital period measured over more than a decade, expressed as phase-lag in seconds; the prediction from GR is indicated by the solid line. The agreement is impressive. This result is the first, albeit indirect, observation of gravitational radiation. We now turn to discuss the sources of intense bursts of gravitational waves which may be detected directly on Earth in the not too distant future.

10.4 Power output of gravitational sources

The sources of gravitational radiation that experimentalists hope to detect directly in the near future are supernovae occurring in our Galaxy. It is essential that the collapse of the stellar core is asymmetric because perfect spherically symmetric motion leaves the external Schwarzschild metric unaffected. Put another way, such motion has no quadrupole or higher moments.

In view of the uncertainties in our knowledge of stellar processes which generate gravitational waves, calculations here are limited to order-of-magnitude estimates of the power output to be expected in a favourable case. The first approximation is to take the quadrupole moment of a body of mass M and size R to be MR^2. Further, the body will be assumed to oscillate at one angular frequency ω. Then the third differential of the reduced quadrupole moment

$$\dddot{I\!\!\!/}_{ij} \approx MR^2\omega^3 \approx Mv^3/R$$

where v is the velocity of material inside the source. The power output (luminosity) is given by substituting this value into eqn (F.6); it yields

$$L \approx \frac{GM^2v^6}{R^2c^5}. \tag{10.18}$$

This can be rewritten as

$$L \approx L_0 \left(\frac{r_0}{R}\right)^2 \left(\frac{v}{c}\right)^6, \tag{10.19}$$

where the factor

$$L_0 = c^5/G = 3.63 \times 10^{52} \text{ W},$$

and r_0 is the Schwarzschild radius of the body. The terms r_0/R and v/c in eqn (10.19) will approach unity for stars collapsing to a neutron star or to a black hole, and so these collapsing stars are potentially important sources of gravitational radiation provided that the collapse is sufficiently asymmetric. In the extreme case the maximum conceivable luminosity reaches the spectacular value of L_0 or 3.63×10^{52} W. At the opposite end of the scale all terrestrial sources are feeble. A massive 1000 t steel rotor with 10 m long arms could be spun so that the tips moved at 1000 m s^{-1}, equalling the speed of sound in steel. The rotor tips would then be on the point of breaking up and the mounting on the verge of shattering under the centrifugal force. The use of eqn (10.18) shows that this monster would only radiate 10^{-30} W of gravitational wave power: this rules out any laboratory equivalent of Hertz's experiment for the production and detection of gravitational waves.

A type II supernova collapse leading to a neutron star occurs on average once every 30 years in our Galaxy. The astrophysics of such events is complicated, so that it is not possible to give more than limits on the parameters of the gravitational radiation to be expected. A burst of frequency between 10 and 10^4 Hz is expected with a duration of several cycles and an energy release of between 10^{-5} Mc^2 and 10^{-2} Mc^2 in gravitational waves from a star core of mass M. The mass M is expected to be similar to that of the pulsars, which average around 1.4 $M_\odot$ for the few cases where measurements have been made. The more asymmetric the collapse, the larger the gravitational energy release becomes. Pulsars are spinning very rapidly so that some asymmetry is likely. As a point of reference it will be assumed that for a supernova collapse the gravitational waves carry off a total energy of 10^{-3} $M_\odot c^2$ in a burst of milliseconds length and at a frequency of about 1 kHz.

Equation (E.8) can be used to convert the energy flux in gravitational waves to the strain in space–time on Earth. When the burst length is $\Delta\tau$ the strain

$$h = h_0 \cos \omega_0 t \qquad |t| < \Delta\tau/2$$

requires an energy flux (recall that $\langle \cos^2 \omega_0 t \rangle$ is 1/2)

$$F(t) = (c^3/32\pi G)\omega_0^2 h_0^2 \qquad |t| < \Delta\tau/2.$$

If the burst originates from a source a distance R from the Earth in which a total mass M is converted to gravitational radiation, we have

$$Mc^2 = 4\pi R^2 \int F\,dt = (c^3/8G)R^2\omega_0^2 h_0^2 \Delta\tau.$$

Thus

$$h_0 = 10^{-18}\,(1\text{ kHz}/\nu)(10\text{ kpc}/R)(M/10^{-3}\,M_\odot)^{1/2}\,(1\text{ ms}/\Delta\tau)^{1/2}. \quad (10.20)$$

The notation used here signifies that the frequency ν is measured in kHz, R is the distance in units of 10 kpc, Mc^2 is the energy in units of $10^{-3}\,M_\odot c^2$ and $\Delta\tau$ is in ms. Thus a type II supernova occurring at the centre of our Galaxy is expected to produce a strain

$$h_0\text{ (Galactic)} \approx 10^{-18}$$

under the assumption that a mass of $10^{-3}\,M_\odot$ is converted to gravitational radiation. Such events occur once every 30 years on average in our Galaxy, and a good proportion are not visible optically because of intervening matter. By extending the region of interest to include the Virgo cluster which contains about 2000 galaxies the rate rises to one every few days, a more tempting prospect. The distance of the Virgo cluster is approximately 15 Mpc, and so the signal is reduced to

$$h_0\text{ (Virgo)} \approx 7 \times 10^{-22}$$

for the conversion of a mass of $10^{-3}\,M_\odot$ to gravitational radiation. Having discussed the parameters of the signals it is time to look at the detectors.

10.5 Detectors

The direct detection of gravitational waves requires the measurement of relative displacements and their changes with time. First a typical interferometric detector will be described.

Figure 10.5 shows a plan view of a large Michelson interferometer in which the three blocks carrying the mirrors M, M_1, and M_2 are freely suspended. When gravitational radiation is incident from above with its polarization axes aligned along the two arms the lengths of these arms will expand and contract in anti-phase. The amplitude of the change in proper path length is given by eqn (10.14) as

$$\Delta l = \frac{h\lambda}{2\pi}\sin\left(\frac{\pi l_0}{\lambda}\right)$$

for a wave of amplitude h and wavelength λ, where l_0 is the unperturbed path length. Such changes in length cause a corresponding oscillation in the fringe pattern. Highly monochromatic laser light is split by the mirror M to travel along the two arms. The beams are reflected at M_1 (M_2), return, recombine coherently at M again, and are detected by a photosensitive device. From the

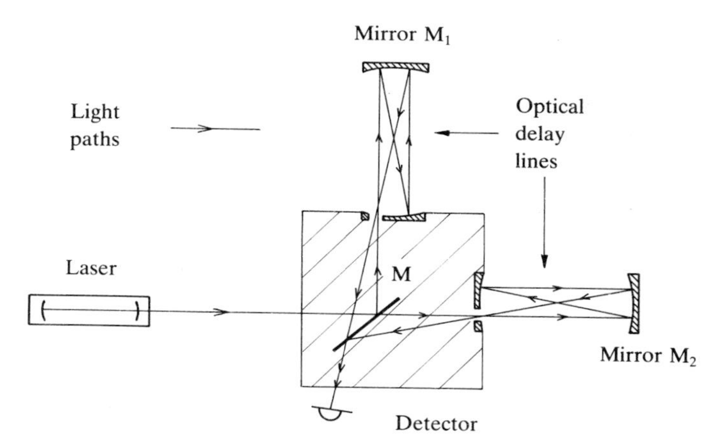

Fig. 10.5 An example of a design for a large interferometer-type detector of gravitational waves.

last equation it is clear that strain is maximal if the optical paths are $\lambda/2$, which for radiation of 1 kHz would be 150 km. Multiple passes in each arm are used to increase the effective optical path length; the scheme shown in Fig. 10.5 with the beam walking across the mirror at successive reflections is known as an optical delay line. Detectors with arms about 20 m long have been built by groups at the Max-Planck-Institut, Garching-bei-München, and at Glasgow University. These and other groups plan to construct detectors with arms about 1 km long and to use about 150 passes. The optical paths are evacuated and the suspension systems are insulated against mechanical vibration by shock absorbers.

The precision of measurement is ultimately restricted by fluctuations in the fringe pattern due to fluctuations in the number of photons detected, because these simulate the effect of optical path changes. Suppose that the wave amplitude in one arm is A, and that in the other arm is $A\exp[\mathrm{i}(2\pi\Delta l/\lambda_{\mathrm{em}})]$ where Δl is the total path difference and λ_{em} is the wavelength of the laser source. Therefore the intensity at the detector is

$$A^2\left[1+\cos\left(\frac{2\pi\Delta l}{\lambda_{\mathrm{em}}}\right)\right]^2 + A^2\sin^2\left(\frac{2\pi\Delta l}{\lambda_{\mathrm{em}}}\right) = 2A^2\left[1+\cos\left(\frac{2\pi\Delta l}{\lambda_{\mathrm{em}}}\right)\right]$$
$$= 4A^2\cos^2\left(\frac{\pi\Delta l}{\lambda_{\mathrm{em}}}\right).$$

The distribution of the number of photons in the fringe pattern is

$$N = N_{\max}\cos^2\left(\frac{\pi\Delta l}{\lambda_{\mathrm{em}}}\right). \tag{10.21}$$

Thus the effect of a change in Δl on N is given by

$$dN = N_{max}\left(\frac{\pi}{\lambda_{em}}\right)\sin\left(\frac{2\pi\Delta l}{\lambda_{em}}\right)d(\Delta l).$$

The statistical error on N is $N^{1/2}$, so that the achievable precision in measuring Δl can be estimated by replacing dN by $N^{1/2}$ in this equation and rewriting it as an expression for $d(\Delta l)$:

$$d(\Delta l) = \frac{\lambda_{em}N^{1/2}}{\pi N_{max}\sin(2\pi\Delta l/\lambda_{em})}.$$

It follows that $d(\Delta l)$ is smallest when $\sin(2\pi\Delta l/\lambda_{em})$ is unity, and then eqn (10.21) gives $N = N_{max}/2 = N_0$, where N_0 is the mean intensity. Substituting these values in the previous equation gives

$$d(\Delta l) = \lambda_{em}/2\pi N_0^{1/2}. \tag{10.22}$$

The value of N_0 depends on the laser power P, the length of time taken for each measurement $\Delta\tau$, and the efficiency ε of detecting photons:

$$N_0 = \Delta\tau\frac{\varepsilon P}{\hbar\omega_{em}}$$

where ω_{em} is the angular frequency of the laser light and $\hbar$ is Planck's constant divided by 2π. Inserting this value for N_0 in eqn (10.22) gives

$$d(\Delta l) = \frac{\lambda_{em}}{2\pi}\left(\frac{\hbar\omega_{em}}{\Delta\tau\varepsilon P}\right)^{1/2}. \tag{10.23}$$

We have seen above that strain will be maximal if the path length l is $\lambda/2$. However, the time that the light remains inside the interferometer arms is then $\lambda/2c$, which means that the measurement time must be equally long or longer. Substituting this limit for $\Delta\tau$ in eqn (10.23) gives

$$d(\Delta l) = \left(\frac{\hbar c\lambda_{em}\nu}{\pi\varepsilon P}\right)^{1/2}.$$

Hence the minimum detectable strain, or the *sensitivity*, is

$$h = \frac{d(\Delta l)}{l} = \frac{d(\Delta l)}{\lambda/2},$$

i.e.

$$h = 2\left(\frac{\hbar\lambda_{em}\nu^3}{c\pi\varepsilon P}\right)^{1/2}. \tag{10.24}$$

An idea of how good this is can be estimated by choosing a laser power of 100 W at 500 nm, $\varepsilon = 0.3$, gravitational waves of frequency 1 kHz and $\Delta\tau$ of a few milliseconds; the estimate is

$$h(\text{min}) \approx 10^{-21}.$$

This implies an optical path length of 150 km, e.g. 150 passes over arms 1 km long. There is therefore a good prospect in the long term of detecting supernovae in our Galaxy or in nearby galaxies. There are plans to improve this limit by 'recycling' the laser light so as to enhance the beam intensities. It is worth noting that, although an interferometer can in principle respond to any frequency, the bandwidth is restricted in practice by the photon storage time to

$$\Delta\nu = \nu^2\Delta\tau < \nu^2(\lambda/2c) = \nu/2.$$

The other type of detector, the resonant bar detector, was invented by Weber (1960) and has been refined by him and others. It consists of a freely suspended bar (antenna) which has some natural frequency ν_0 for acoustic longitudinal vibrations. An incident burst of gravitational radiation with favourable polarization and frequency will excite the bar to oscillate and the oscillations will continue long after the burst has ended. This 'ringing' can be detected by transducers connected mechanically to the bar. The fundamental frequency for longitudinal standing waves on a bar of length l is

$$\nu_0 = \frac{v_s}{2l}$$

where v_s is the speed of sound in the material. For the aluminium alloy 5056 at low temperature v_s is 5.5×10^3 m s^{-1} so that it requires a bar of length 3 m to resonate at 1 kHz, an appropriate frequency for detecting a burst from a type II supernova. As we show next, the gravitational waves from likely sources are only detectable if the bar is cooled to liquid helium temperatures. The energy deposited by a gravitational wave in a bar of mass M and length l is

$$E_g \approx Mv^2/2,$$

where the mean velocity v within the bar is ωhl for a wave of strain amplitude h and angular frequency ω. Then

$$E_g \approx M\omega^2h^2l^2/2. \quad (10.25)$$

For a type II supernova at the centre of our Galaxy $h \approx 10^{-18}$ and so the energy deposited in a bar of mass 2 t and length 3 m would be

$$E_g \approx 10^{-25} \text{ J}.$$

This is to be compared with the thermal energy in any mode of oscillation at 4 K:

$$E_t \approx kT \approx 10^{-23} \text{ J}.$$

On the face of it the thermal noise appears to dominate. What saves the day is the fact that many materials are good resonators. For these there is only weak coupling between the various modes of internal motion so that natural oscillations can continue for a long time; not only this, but the thermal energy in any mode will change with equal slowness. A relatively quick measurement of the oscillations would only be sensitive to the *fluctuations* in the thermal energy E_t over the measurement interval. The quality of a resonator is given by the Q factor

$$Q = \omega_0/2\gamma.$$

Here ω_0 is the natural frequency and γ is the time constant for the decay of oscillations of amplitude $A(t)$:

$$A(t) = A(0)\exp(-\gamma t)\cos\omega t.$$

The aluminium alloy 5056 can have $Q \approx 10^7$, so that at 10^3 Hz γ is around $3 \times 10^{-4}\,\mathrm{s}^{-1}$. A measurement lasting 1 s would be subject to a thermal fluctuation

$$E_t\gamma \approx 3 \times 10^{-27}\ \mathrm{J}.$$

This reduction in thermal noise make it realistic to attempt to detect gravitational waves with a suitably cooled massive bar. Figure 10.6 shows a modern antenna built by the Roma group at CERN (Amaldi *et al.* 1986). The solid cylinder of 5056 aluminium is 60 cm in diameter and 3 m long, weighs 2.27 t, and is suspended *in vacuo* by a titanium cable wrapped under the centre of gravity. The overall damping of the suspension system against external mechanical shock is approximately 10^{-13} and the antenna is in an enclosure at 4.2 K. Its resonant frequency is 916 Hz and $Q \approx 4.4 \times 10^6$. A sampling time of 0.3 s is used and the relaxation time $1/\gamma$ is 25 min. Transducers and amplifi-

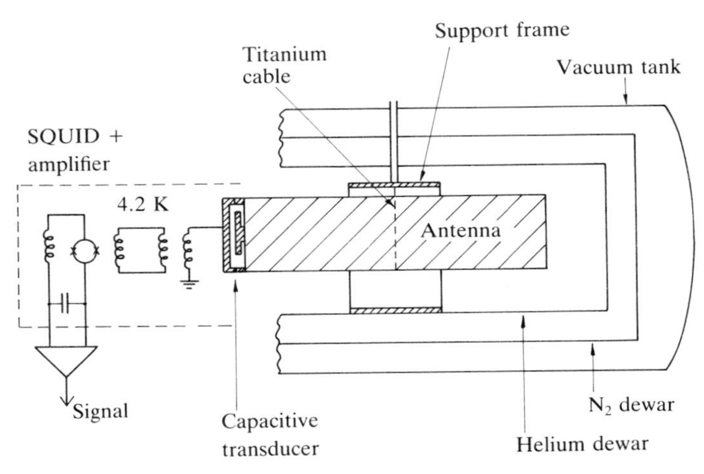

Fig. 10.6 A schematic diagram of the Roma gravitational wave antenna and transducer. (After E. Amaldi *et al.* (1986), with kind permission of the authors and Societa Italiana di Fisica.)

cation of electronic signals require careful design to achieve low noise combined with adequate signal transfer. Amaldi *et al.* use a capacitor to detect the motion and a superconducting transformer to couple the induced signals to a d.c. superconducting quantum interference device (SQUID).

Simple resonant bar devices face a limit of sensitivity imposed by quantum theory. Present methods, as described above, are not capable of detecting a signal which produces an energy change in the mode of vibration of less than $\hbar\omega_0$, i.e. the energy must change by one or more phonons. Thus

$$E_g > \hbar\omega_0.$$

Using eqn (10.25) this becomes

$$Ml^2\omega_0{}^2h^2 > \hbar\omega_0.$$

The quantum limit on detectable strain is therefore

$$h(\text{min}) \approx (\hbar/l^2\omega_0 M)^{1/2} \approx 10^{-21}$$

for a modern antenna. The sensitivity of resonant bars is thus comparable with that of interferometers. In the longer term ways have been proposed to bypass the quantum limit. It has been noted that the quantum limit has its origin in the following expression of the uncertainty principle. Conjugate quantities such as the phase and amplitude of an oscillation cannot be simultaneously measured with arbitrary precision because their operators do not commute. However, if only *one* variable, such as the amplitude, is measured it can be measured with arbitrary precision. Various ways of using this loophole are under current investigation.

To summarize, it seems that detectors which will go into *continuous* operation in the next few years have a strong chance of detecting a type II supernova in our Galaxy provided that the collapse is asymmetric to the degree now thought likely.

11
Cosmology

To begin with we review the basic facts established by astronomers and astrophysicists, who have shown that the Universe is expanding and is over ten billion years old (Section 11.1). A simple model of the Universe which provides a framework for discussing its development is to take the Universe to be isotropic and homogeneous. In this approximation the structure of space–time is described by the Robertson–Walker metric (Section 11.2). The dynamics of how such a universe develops is obtained by assuming matter is in the form of a homogeneous ideal fluid and then applying Einstein's equation (Section 11.3). The resulting equations provide a framework for discussing plausible histories and futures for the Universe (Section 11.4). In Section 11.5 the existence of the cosmic background radiation (CBR) and the abundance of light nuclei are discussed. Both are features that are very simply explained if the Universe expanded from a very small initial volume, perhaps an initial singularity. Olbers' paradox is reviewed in Section 11.6. The isotropy of the CBR is hard to understand; one tentative explanation is that the early Universe went through a phase of spectacular inflation (Section 11.7).

11.1 Basic observations

The matter in the visible Universe is concentrated in galaxies, of which our own is fairly typical: its stars populate a pancake-shaped region of diameter 30 kpc and about 1 kpc thick. It contains about 10^{11} stars whose average mass is comparable with that of our own Sun. The galaxies form clusters and superclusters extending over tens to hundreds of megaparsecs. One large cluster, the Virgo cluster containing about 2000 galaxies, is about 15 Mpc from the Earth: it forms the core of the local supercluster that includes our own Galaxy and the local group of some 24 galaxies. Structure is visible up to the largest distances studied; superclusters and clusters form threads and sheets that are separated by huge voids where galaxies are very rare. These voids can by 100 Mpc across, such as, for example, the Böotes void.

A feature obvious to anyone looking at the night sky is that it is not uniformly bright. If the Universe were infinite in extent and the galaxies we see now had existed eternally then the luminous intensity would be the same in all directions. Evidently the Universe is finite in extent or its features are finite in duration. Another fundamental observation made by Penzias and Wilson (1965) is that the Earth is bathed in microwave radiation not originating from stars of galaxies. Figure 11.1 shows a compilation of recent measurements of the microwave spectrum taken from Matsumo *et al.* (1988). Over the wavelength range from 1 mm to 10 cm the spectrum is consistent with that of black-body radiation at 2.74 K. Outside this spectral window the emission from our own Galaxy masks the tails of the black-body spectrum; at shorter wavelengths the emissions from interstellar dust (ISD) are seen to be important. The black-body spectrum is known as the cosmic background radiation (CBR) and cannot be explained by emission from any galactic material. There is a small anisotropy in the temperature of CBR, which varies with direction according to the formula

$$T(\theta) = T_0 + \Delta T \cos \theta$$

where the pole ($\theta = 0$) has Galactic latitude $54° \pm 10°$ and longitude $245° \pm 15°$. This anisotropy is readily explained if the Earth is in motion relative to the frame of the CBR; the induced Doppler shift of the radiation then produces the observed cosine dependence on angle with respect to the Earth's velocity vector. With respect to the frame in which the radiation is isotropic the Earth's velocity is found to be 400 km s^{-1}. When the measurements on the CBR are transformed to this frame it is found that the temperature across the sky is constant to better than one part in 10^4. The current energy density of the CBR is given by $\varepsilon_{r0} = aT_0{}^4$ where a is the radiation density constant (7.56×10^{-16} J m^{-3} K^{-4}) and $\varepsilon_{r0} = 4.3 \times 10^{-14}$ J m^{-3}. Two important infer-

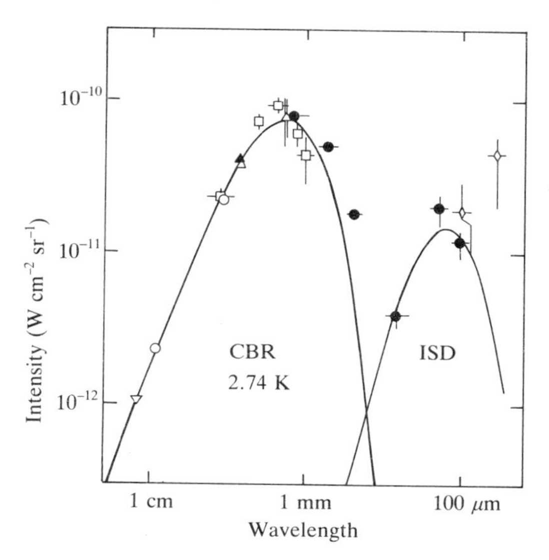

Fig. 11.1 The spectrum of the cosmic background radiation with the black-body spectrum (solid line) for 2.74 K. (Adapted from Matsumoto *et al.* 1988; courtesy of Professor Lange and the *Astrophysical Journal*, published by the University of Chicago Press.)

ences can be made directly from the existence of the CBR. Firstly, this radiation identifies a preferred reference frame and so contradicts the inference often made from SR that no frame is preferred. Secondly, if this radiation is a relic from the remote past it shows that in its early life the Universe was isotropic and homogeneous, unlike today.

Hubble (1929) demonstrated that the spectral lines emitted from remote sources appear systematically shifted toward the red end of the spectrum. This effect is *now* interpreted as being due to the expansion of the Universe and the consequent lengthening of the wavelength of electromagnetic radiation during the time that it is travelling towards Earth. We define the parameter.

$$z = \frac{\lambda - \lambda_e}{\lambda_e}$$

where λ_e is the wavelength of a spectral line observed from a source on Earth and λ is the wavelength of the same line in the spectrum of a distant Galaxy. Hubble discovered that the red shift z is proportional to our distance d from the source:

$$z = Hd/c \tag{11.1}$$

where H is known as Hubble's constant. Recent studies have shown that locally the gravitational attraction of nearby clusters and superclusters distorts the uniform expansion. Attempts to use more distant sources meet with difficulties because of the lack of reliable distance indicators; the ideal would be 'standard candles', i.e. sources of constant intensity in the visible or radio spectrum. It is expected that when the Hubble space telescope is put into Earth orbit from the Space Shuttle that this telescope will be able to detect supernovae in remoter galaxies and use these as standard candles. For the present Hubble's constant H_0 for our epoch is only known to modest precision:

$$H_0 = (50\text{–}100)\ \text{km s}^{-1}\ \text{Mpc}^{-1}.$$

It is significant that an expansion velocity proportional to distance can be imposed everywhere in space. If the velocity were proportional to any other power of the distance this universal behaviour would not be possible (see Question 11.1). This suggests that if we could look at the Universe from any other galaxy, the view would be very much the same as that from Earth. As remarked above, galactic clustering must produce local distortions on a uniform expansion. Attempts to determine this local motion for the Earth relative to the distant galaxies give a velocity for the Earth with a direction and magnitude which roughly match the Earth's velocity relative to the CBR. It is therefore likely that the frame in which the Hubble motion of distant galaxies is isotropic and the frame in which the CBR is isotropic are one and the same. This will be the view taken below.

Radioactive dating of the oldest rocks found on earth, of meteorites, and of lunar material gives ages of around 4.5×10^9 years. Other measurements of the lifetime of the Galaxy itself have been made using the observed relative abundances of pairs of related radioactive species (^{235}U and ^{238}U). Such heavy nuclides are produced when stars complete their cycle of thermonuclear burning and subsequently explode as supernovae. The initial relative abundance of pairs of heavy nuclear species from this source in the interstellar medium can be calculated with fair accuracy. ^{238}U has a half-life of 4.5×10^9 years and ^{235}U a half-life of 7.1×10^8 years, and so the relative abundance changes with time in a well-defined way. The measured current relative abundance implies a galactic lifetime of around 15×10^9 years. This sets a *lower* limit on the age of the Universe. Models of stellar evolution are used to make other estimates of the lifetime of the oldest stars in the Galaxy, which also come out at around 15×10^9 years.

All basic observations of the Universe are consistent with the *cosmological principle*, namely that the view from Earth is a typical view of the Universe. We shall adopt this principle here because without some such principle it would be impossible to draw inferences about the history of the Universe. From the evidence of the CBR we can say that the Universe was once isotropic and homogeneous to a high degree. At present the distribution of matter shows structures on all scales so far explored with the largest structures being random in distribution. Details of possible mechanisms by which the structures could have evolved can be found in Silk (1980).

If the expansion of the Universe is extrapolated back in time, then naively we would expect that all points in space would be superposed at some initial instant. This is an initial singularity of both the mathematical and the physical variety. Assuming that the Universe expanded at the same rate as at present, the singularity would have occurred $(1–2) \times 10^{10}$ years ago. An important feature of the early Universe would have been its high temperature. The peak wavelength of the CBR would have been shorter by the ratio of the size of the Universe then to its current size, and so the CBR temperature would have been higher. The Universe would have commenced with a 'hot big bang'.

11.2 The Robertson–Walker metric

The simple model for the Universe which we shall make much use of is to take it to be homogeneous, isotropic, and with matter uniformly distributed at the mean density of the Universe. Instead of galaxies and stars this model has a uniform perfect fluid filling all space. It seems that this was how the Universe appeared when the CBR was emitted, and the galaxies now present have developed since that epoch. The average motion of a galaxy is expected to follow that of a particle of the ideal fluid. Our first step is to deduce the metric of this model universe.

The development of this universe will be the same everywhere. We can imagine placing clocks at rest with respect to the cosmic fluid and setting them to read the same reference time when the fluid density and temperature reach agreed values. Then using these synchronized clocks the physical state of the universe will depend on time in the same way everywhere. This time will be called the *cosmic* time. Suppose now that a three-dimensional slice (hypersurface) is taken through space–time at cosmic time t. This hypersurface will also be isotropic and homogeneous. Therefore the Gaussian curvature of all geodesic surfaces in this hypersurface will have *one* value which depends on cosmic time:

$$K(t) = k/R^2(t) \tag{11.2}$$

where R^2 gives the magnitude and k the sign ($+1$, , 0, 1) of this curvature.

A positively curved space with $k = +1$ is the three-dimensional analogue of a spherical surface, i.e. a hypersphere. It can be embedded in a four-dimensional Euclidean space (E_4) just as a two-dimensional spherical surface is embedded in Euclidean three-space (E_3). Let x, y, z, and w be the Cartesian coordinates in E_4 with x, y, and z being the usual spatial coordinates of an E_3. Then the hypersurface has the equation

$$x^2 + y^2 + z^2 + w^2 = R^2(t),$$

i.e.

$$r^2 + w^2 = R^2(t),$$

where r, θ, and φ are spherical polar coordinates in E_3. By differentiation we obtain

$$w\,dw = r\,dr, \qquad \text{and} \qquad w^2\,dw^2 = r^2\,dr^2.$$

Then

$$dw^2 = \frac{r^2\,dr^2}{R^2 - r^2}.$$

The separation of nearby points on the hypersurface is given by

$$dl^2 = dr^2 + r^2\,d\Omega^2 + dw^2.$$

Eliminating dw^2 we obtain the metric equation of the hypersurface:

$$\begin{aligned} dl^2 &= \frac{R^2\,dr^2}{R^2 - r^2} + r^2\,d\theta^2 + r^2 \sin^2\theta\,d\varphi^2 \\ &= \frac{R^2\,dr^2}{R^2 - r^2} + r^2\,d\Omega^2. \end{aligned}$$

A change in angle θ produces a displacement $r\,d\theta$, while a change in r in any direction gives a displacement of $R\,dr/(R^2 - r^2)^{1/2}$. These features show that the three-dimensional hypersurface is curved and isotropic. The choice of positive curvature means that it is a hypersphere. Note that the metric equation is the same for all points on the hypersphere, and so this space is homogeneous also. When the variable $\sigma = r/R$ is used, the metric equation reduces to

$$dl^2 = R^2(t)\left(\frac{d\sigma^2}{1-\sigma^2} + \sigma^2\, d\Omega^2\right).$$

Now we can write the complete metric equation incorporating time:

$$ds^2 = c^2\, dt^2 - R^2(t)\left(\frac{d\sigma^2}{1-\sigma^2} + \sigma^2\, d\Omega^2\right).$$

Any point moving so that the spatial coordinates σ, θ, and φ remain constant will follow a particularly simple world line. First of all

$$\frac{dx^i}{ds} = \frac{d^2x^i}{ds^2} = 0 \qquad \text{and} \qquad \Gamma^i_{00} = 0$$

for spatial coordinates along this world line. Such a world line therefore automatically satisfies the geodesic equation

$$\frac{d^2x^i}{ds^2} + \Gamma^i_{\mu\nu}\left(\frac{dx^\mu}{ds}\right)\left(\frac{dx^\nu}{ds}\right) = 0.$$

In addition the metric equation reduces to

$$ds^2 = c^2\, dt^2$$

which means that the cosmic time t is the same as the proper time τ. The world line is therefore in free fall in the cosmic fluid. For this reason σ, θ, and φ are known as *comoving coordinates.* A frame with constant comoving coordinates is called a comoving frame. Any real galaxy or cluster on which the net attraction of local matter is small is comoving.

An observer in a comoving frame would see the remote galaxies moving isotropically, i.e. radially, just as Hubble discovered. It follows from the strong equivalence principle that because a comoving frame is in free fall the laws of physics in such a frame should satisfy the postulates of SR. In the language of SR the comoving frame is an inertial frame. Other inertial frames can be obtained from the comoving frame by boosting to any frame with constant relative velocity. These are equally good inertial frames for SR (or for Newtonian mechanics if the velocity is small). However, in the Robertson–Walker model the comoving frame is unique because it is the only frame in which the CBR is seen as isotropic. Boosting to another frame gives the CBR a Doppler shift which renders it anisotropic. Consequently the comoving frame is a *preferred* frame, something abhorrent to both classical Newtonian and SR views. The *preferred* frame is particular to a locality in the Universe: comoving frames at different locations have different velocities and accelerations. The Robertson–Walker view of the Universe is fully consistent with GR, once the equivalence between comoving frames and frames in free fall is noted.

Mach's principle in one form states that the distribution of matter in the Universe is responsible for inertial effects. This has a natural explanation within the framework of the Robertson–Walker model. Distant matter in the form of the distant galaxies provides a reference frame that is simply the comoving frame. This is why, for example, a Foucault pendulum at the pole swings in a plane that is fixed with respect to the distant galaxies.

So far we have imagined that the hypersurface at constant time had positive curvature. More generally the curvature k could be negative or zero. The metric equation which covers all these cases is

$$ds^2 = c^2\,dt^2 - R^2(t)\left(\frac{d\sigma^2}{1 - k\sigma^2} + \sigma^2\,d\Omega^2\right) \tag{11.3}$$

where k is taken from eqn (11.2). $R^2(t)$ is the magnitude of curvature, and so we call $R(t)$ the *scale factor* of the Universe; quite clearly it is the hyper-radius of space only when k is $+1$. Equation (11.3) is the metric equation discovered independently by Robertson (1936) and Walker (1936) for isotropic and homogeneous space–time.

Taking our galaxy as origin and aligning the coordinates appropriately, we can make the comoving coordinates of any other galaxy $(\sigma, 0, 0)$. The distance across the hypersurface to this second galaxy is

$$d = R(t)\int_0^{\sigma} \frac{d\sigma}{(1 - k\sigma^2)^{1/2}} \tag{11.4}$$

which gives contrasting results according to the sign of k. Let us take $k = +1$ first. Then

$$d = R(t)\sin^{-1}\sigma.$$

If a sphere centred at the Earth is drawn through the second galaxy the area of the sphere is

$$A = 4\pi R^2\sigma^2 = 4\pi R\,\sin^2(d/R).$$

As d increases from zero to $\pi R/2$ the area of the sphere increases steadily. Then as d increases further to πR the area of the sphere falls to zero: by analogy with the case of the two-dimensional sphere (Section 3.5) we are at the antipodes. Finally when d reaches $2\pi R$, σ is zero and we are back at the origin. A hypersurface with positive curvature is a closed but unbounded surface. On the other hand, when $k = -1$ we have

$$d = R(t)\sinh^{-1}\sigma.$$

The area of the sphere through the other galaxy is

$$A = 4\pi R^2\sigma^2 = 4\pi R^2\,\sinh^2(d/R)$$

which grows indefinitely as σ and d increase together. This type of universe is unbounded and open. For the remaining case that $k = 0$ the hypersurface is the flat Euclidean space of classical mechanics: it is open and unbounded. In a flat universe $d = R(t)\sigma$.

Measurements of the Hubble constant give information on the behaviour of R. The wavelengths appearing in red-shift formulae depend on R; as R increases so does the wavelength λ. Thus

$$z = \frac{\lambda - \lambda_e}{\lambda_e}$$

$$= \frac{(R - R_e)}{R_e}$$

$$= \frac{\dot{R}\Delta t}{R}$$

where Δt is the travel time between the source galaxy and Earth. Comparison with eqn (11.1) shows that if d/c is interpreted as the travel time Δt

$$H = \dot{R}/R.$$

In order to refine this result we make a series expansion for R at the time of emission ($R_e \equiv R(t_e)$) in terms of $\Delta t = t - t_e$:

$$R_e = R - \Delta t\dot{R} + \Delta t^2\ddot{R}/2 + \cdots$$

where R, $\dot{R}$, and $\ddot{R}$ are values measured in the frame of the observer on Earth. Thus

$$R_e = R\,\{1 - \Delta tH - (\Delta t)^2 qH^2/2 + \cdots\}$$

where

$$H = \dot{R}/R \tag{11.5}$$

and

$$q = -\ddot{R}/RH^2 \tag{11.6}$$

Use of the subscript 0 signifies quantities measured at the current epoch. q is called the *deceleration parameter* and would be positive if the expansion were slowing down. Using this expansion

$$z = R/R_e - 1$$

$$= \{1 - \Delta tH - (\Delta t)^2 H^2 q/2 + \cdots\}^{-1} - 1$$

$$= H\Delta t + (1 + q/2)H^2(\Delta t)^2 + \cdots.$$

As mentioned earlier H_0 is 50–100 km s^{-1} Mpc^{-1} whilc q_0 is less well determined: $-1.3 < q_0 < 2.0$.

The expansion of the Universe has affected the CBR by red shifting it; thus at earlier epochs the temperature of the radiation would have been much larger. It is easy to deduce how the temperature depends on R. Assume that on emission the spectrum had the black-body distribution for a temperature T:

$$\mathrm{d}N = \frac{8\pi\nu^2\,\mathrm{d}\nu\,V}{c^3\{\exp(h\nu/kT) - 1\}}$$

in a comoving volume V. During expansion this volume has changed to

$$V' = (R'/R)^3 V$$

but contains the same number of photons $\mathrm{d}N' = \mathrm{d}N$. The changes of wavelength and frequency during expansion are given by

$$\lambda' = (R'/R)\lambda \qquad \text{and} \qquad \nu' = (R/R')\nu.$$

Therefore the equation for $\mathrm{d}N$ can be rewritten in terms of primed quantities as

$$\mathrm{d}N' = \frac{8\pi(\nu')^2 V'\,\mathrm{d}\nu'}{c^3\{\exp(h\nu'/kT') - 1\}}$$

where $T' = T(R/R')$. We see that the spectrum remains that of a black body in shape and intensity. The earlier temperature T' was higher and the temperature has fallen in proportion to the expansion. Consequently the radiation (and the Universe) were hotter in the past and would have reached unimaginable values if the well-established expansion of the Universe started with an initial singularity.

How the Universe expanded from the initial singularity is easiest to picture for a Universe of positive curvature. The Universe at each moment of cosmic time would be a hypersphere of radius R in four-dimensional space (E_4). It would expand like a balloon, with the galaxies (once created), being analogous to dots marked out on the balloon's surface. This balloon analogy brings out an important feature of the expanding Universe, namely each point in the current Universe originated at the initial singularity and they have all moved away from it.

The development and interaction of the Universe is portrayed quite effectively using comoving coordinates. This is shown in Fig. 11.2. Only one of the spatial coordinates is drawn in; σ is the natural choice. When there are no local inhomogeneities matter will follow world lines at constant σ (e.g. the vertical lines from A, B, and C). These would also be world lines for galaxies after their formation. If the Universe did *not* begin, then each world line would continue indefinitely into the past. Figure 11.2 shows the case that the Universe began ($t = 0$) with $R = 0$; the wobbly line indicates the initial singularity. Also shown in Fig. 11.2 are the past light-cones at times t_1 and t_2 for B. The equation for light travelling radially is

$$0 = c^2\,\mathrm{d}t^2 - \frac{R^2\,\mathrm{d}\sigma^2}{1 - k\sigma^2};$$

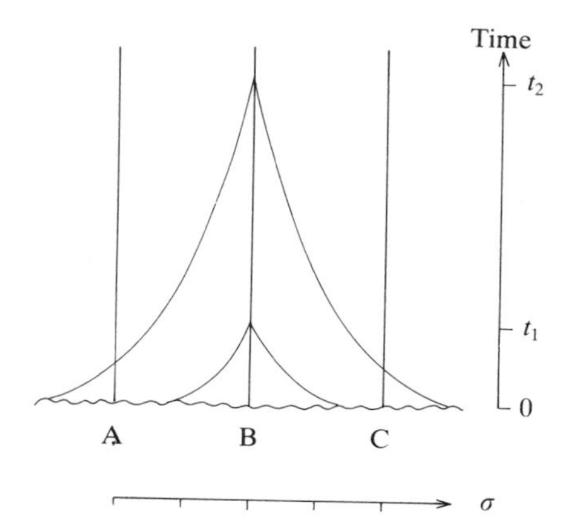

Fig. 11.2 The development of the Universe from a singularity. The past light cones are drawn for B at two epochs t_1 and t_2.

therefore the velocity of light at $\sigma = 0$ is

$$\frac{\mathrm{d}\sigma}{\mathrm{d}t} = \frac{c}{R}.$$

Consequently the light cones in an expanding Universe are concave up. Events lying within the past light cone at t_1 can influence B at time t_1. The boundary of this region is called the *particle horizon* for B at that time. From this observation we can deduce that at time t_1, A and C were not inside the horizon for B. Equally at time t_2, A and C have entered the horizon of B. Events early in the lives of A and C therefore affect B at time t_2 but not at time t_1. B is said to be in causal contact with A and C at time t_2, whilst at time t_1 it was not yet in causal contact with them.

Another horizon is commonly defined called the *event horizon*. This is the surface on which lie the most distant events occurring at a given cosmic time which would ever be observable from a given location. For instance, the furthest events now occurring that will ever be visible from the Earth constitute our event horizon.

It is worth examining further the situation at time $t = 0$. The separation AB across a hypersurface at constant time t is

$$d = R(t) \begin{cases} \sinh^{-1} \sigma & \text{if } k = -1 \\ \sigma & \text{if } k = 0 \\ \sin^{-1} \sigma & \text{if } k = +1, \end{cases}$$

which for $R(0) = 0$ would be zero at $t = 0$ whatever the value of k. This suggests a dilemma. How is it that A and B are not in causal contact at $t = 0$?

The answer is that, although their separation is *instantaneously* zero across a hypersurface at constant time, this does not mean that light could travel between these points. Light cones are only tangential to the σ axis at $t = 0$ but develop thereafter as shown in Fig. 11.2.

We on Earth might expect to see the origin of the Universe (in all directions!) but because radiation has scattered numerous times since then the image is lost. The CBR preserves a picture of the Universe when radiation was last in equilibrium with matter. That part of the Universe which is visible to us through the CBR was apparently homogeneous in temperature to better than one part in 10^4, but this does not guarantee anything about the regions that we do not see. We choose to assume that we have an average view of the Universe (the cosmological principle). It is quite conceivable that future tests of the models of the Universe which we build up from this assumption may reveal inconsistencies and it would become necessary to revise it. For the present it is a good working hypothesis.

11.3 Cosmic dynamics

The parameters of supreme interest for the physicist are the sign of curvature and the magnitude of the scale parameter of the Universe. Einstein's equation provides the dynamical link between the metric and the matter content of the Universe and so offers the hope that we can calculate the time dependence of $R(t)$. Once we know this we may be able to infer the life story and future of the Universe. The stress-energy tensor of the perfect fluid used to mimic the matter content of the Universe is

$$T_{\mu\nu} = (p/c^2 + \rho)v_\mu v_\nu - p g_{\mu\nu} \tag{11.7}$$

where p is the pressure, ρ the rest density, and v the fluid velocity. We now proceed to set down the component pieces of Einstein's equation. Using the Robertson–Walker metric (eqn (11.3)),

$$g_{00} = g^{00} = 1 \qquad g_{11} = \frac{1}{g^{11}} = \frac{-R^2}{1 - k\sigma^2}$$

$$g_{22} = \frac{1}{g^{22}} = -R^2\sigma^2 \qquad g_{33} = \frac{1}{g^{33}} = -R^2\sigma^2 \sin^2\theta.$$

In a comoving frame $v = (c, 0, 0, 0)$ and

$$T_{00} = \rho c^2 \qquad T_{11} = \frac{pR^2}{1 - k\sigma^2}$$

$$T_{22} = pR^2\sigma^2 \qquad T_{33} = pR^2\sigma^2 \sin^2\theta.$$

The metric connections are evaluated with the help of eqn (6.5); for example,

$$\Gamma^1_{01} = \Gamma^1_{10} = g^{11}\Gamma_{110} = \frac{1}{2}g^{11}g_{11,0}$$

$$= \frac{1}{2}\left(-\frac{1-k\sigma^2}{R^2}\right)\left\{-\frac{2R\dot{R}}{c(1-k\sigma)^2}\right\}$$

$$= \frac{\dot{R}}{Rc}.$$

Equation (7.3) gives the components of the Riemann tensor: for example

$$R^1_{010} = -\Gamma^1_{01,0} - \Gamma^1_{10}\Gamma^1_{01}$$

$$= \left(\frac{\dot{R}^2}{R^2c^2} - \frac{\ddot{R}}{c^2R}\right) - \frac{\dot{R}^2}{c^2R^2}$$

$$= \frac{-\ddot{R}}{c^2R}.$$

Similarly $R^2_{020} = R^3_{030} = -\ddot{R}/c^2R$ and R^0_{000} is identically zero. Using eqn (7.14) the 00 component of the Ricci tensor is

$$R_{00} = R^0_{000} + R^1_{010} + R^2_{020} + R^3_{030} = -3\ddot{R}/c^2R.$$

Its only other non-zero components are

$$R_{11} = \frac{T}{1-k\sigma^2} \qquad R_{22} = T\sigma^2 \qquad R_{33} = R_{22}\sin^2\theta$$

where

$$T = 2k + R\ddot{R}/c^2 + 2\dot{R}^2/c^2.$$

Thus the Ricci scalar is given by

$$R(\text{Ricci scalar}) = g^{\mu\nu}R_{\mu\nu} = -6S/R^2$$

where

$$S = k + R\ddot{R}/c^2 + \dot{R}^2/c^2.$$

Finally the Einstein tensor (eqn (7.15)) is given by

$$G_{\mu\nu} = R_{\mu\nu} - \tfrac{1}{2}g_{\mu\nu}R(\text{Ricci scalar})$$

for which the first two non-vanishing components are

$$G_{00} = 3\dot{R}^2/R^2c^2 + 3k/R^2$$

$$G_{11} = -\frac{k + 2R\ddot{R}/c^2 + \dot{R}^2/c^2}{1-k\sigma^2}.$$

The corresponding 00 and 11 components of the Einstein equation (7.17) are thus

$$3\dot{R}^2/R^2 + 3kc^2/R^2 - c^2\Lambda = 8\pi G\rho \tag{11.8}$$

and

$$-2\ddot{R}/R - \dot{R}^2/R^2 - kc^2/R^2 + c^2\Lambda = 8\pi Gp/c^2 \tag{11.9}$$

and where Λ is the cosmological constant. Other components of the Einstein equation duplicate these results and add no further dynamical information. Equations (11.8) and (11.9) were discovered by Friedmann (1922) for the case that $p = 0$, and were generalized by Le Maître (1927). There remains the question of how the pressure depends on the density: in order to complete the description an equation of state $p = p(\rho)$ is needed.

At the present epoch the pressure p is small and so can be set to zero; eqn (11.9) becomes

$$-2\ddot{R}/R - \dot{R}^2/R^2 - kc^2/R^2 + c^2\Lambda = 0 \tag{11.10}$$

Then if the cosmological constant also vanishes eqns (11.8) and (11.10) reduce to

$$3\dot{R}^2/R^2 + 3kc^2/R^2 = 8\pi G\rho \tag{11.11}$$

$$2\ddot{R}/R + \dot{R}^2/R^2 + kc^2/R^2 = 0. \tag{11.12}$$

We shall show now that these last two equations have a simple interpretation in the Newtonian limit.

Figure 11.3 shows a comoving spherical surface of radius σR in the cosmic fluid. If curvature and local velocities are small then Newtonian mechanics will apply to the motion of material at the surface. Its velocity is $\sigma\dot{R}$ outward and so the kinetic energy per unit mass is $\sigma^2\dot{R}^2/2$. The gravitational energy of this elementary mass due to the mass contained inside the sphere is $-(4\pi/3)G\rho R^2\sigma^2$. All material outside the sphere produces zero net effect, in

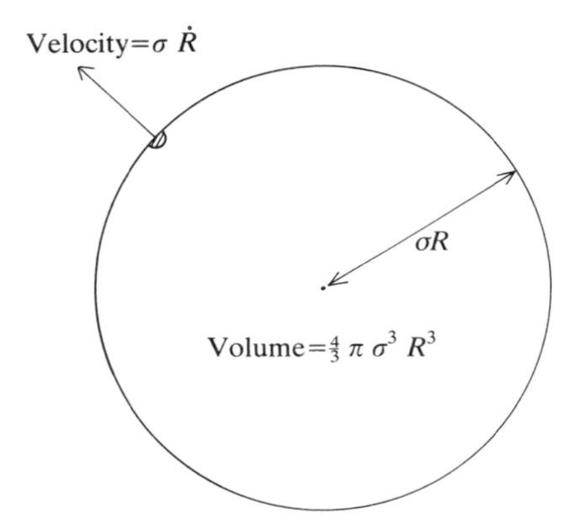

Fig. 11.3 A comoving sphere inside the isotropic homogeneous cosmic fluid.

Table 11.1 The relationship between net energy and curvature in homogeneous, isotropic universes with vanishing cosmological constant

k	*Curvature*	*Net energy*
$+1$	Positive	Negative
0	Flat	Zero
-1	Negative	Positive

the same way that the electrical potential within a hole inside a conductor is uniform. Thus the total energy of this elementary mass is

$$\frac{\sigma^2 \dot{R}^2}{2} - \frac{4\pi}{3} G\rho R^2 \sigma^2 = \text{constant.}$$

If we choose to write the constant as $-\sigma^2 c^2 k/2$, then after some rearrangement we obtain

$$3\dot{R}^2/R^2 + 3kc^2/R^2 = 8\pi G\rho$$

which reproduces eqn (11.11). The parameter k is now seen to have a two-fold significance: first k is the sign of curvature of the hypersurface (constant t); second $-k$ is the sign of the net energy of matter in the Universe. Table 11.1 summarizes this simple situation existing when $\Lambda = p = 0$. Another simple result follows if we multiply eqn (11.11) by R^3 and then differentiate with respect to t:

$$6\ddot{R}\dot{R}R + 3\dot{R}^3 + 3kc^2\dot{R} = \frac{\mathrm{d}}{\mathrm{d}t}(8\pi G\rho R^3).$$

From eqn (11.12) we see that the left-hand side vanishes, and so

$$\frac{\mathrm{d}}{\mathrm{d}t}\left(\frac{4\pi G\rho R^3}{3}\right) = 0,$$

which means that the mass $4\pi\rho R^3/3$ within the comoving volume does not change. This is called the *continuity* equation. In summary, eqns (11.11) and (11.12) are equivalent in the classical limit to the conservation of energy equation and the continuity equation.

11.4 The past and future of model universes

Equations (11.8) and (11.10) tell how the idealized homogeneous isotropic universe would develop at the present epoch when the pressure p is negligible (dust universe). First we rewrite them in terms of measurable quantities H and

q using eqns (11.5) and (11.6). Then after some rearrangement,

$$3kc^2/R^2 = 8\pi G\rho - 3H^2 + c^2\Lambda \tag{11.13}$$

$$kc^2/R^2 = (2q - 1)H^2 + c^2\Lambda. \tag{11.14}$$

Unfortunately H, q, and ρ are not yet well measured. From the first equation we see that

$$k > 0 \text{ if } \rho > \frac{3H^2 - c^2\Lambda}{8\pi G}.$$

The *critical* or *closure density* ρ_c, at which $k = 0$, is given by

$$\rho_c = \frac{3H^2 - c^2\Lambda}{8\pi G}. \tag{11.15}$$

If, in addition, $\Lambda = 0$ (which will be our viewpoint for almost all of this section),

$$\rho_c = 3H^2/8\pi G. \tag{11.16}$$

In the current epoch

$$\rho_c = (0.5\text{–}2.0) \times 10^{-26}\ \text{kg m}^{-3},$$

taking into account the uncertainty in H_0.

The density of radiating matter in galaxies is about 10^{-29} kg m^{-3}, assuming an average mass-to-luminosity ratio the same as that of the Sun. This is not the whole story because there are good dynamical reasons to suppose that much more matter is present which radiates little, if at all. One such line of reasoning starts from determinations of the velocity of gas clouds circulating around the periphery of galaxies. These determinations are based on measurements of the red shift of the 21 cm hydrogen emission line. By subtracting the mean red shift for the parent galaxy, an estimate is obtained of the velocity of each local region relative to the galaxy. Data collected by Rubin and Ford (1980), and shown in Fig. 11.4, illustrate how the tangential velocities v change as a function of the distance r from the galactic centre. If the gas clouds are moving in circular orbits, then

$$v^2 = GM(r)/r$$

where $M(r)$ is the mass of galactic material inside a radius r. We would expect v to fall like $1/r^{1/2}$ in the outer reaches, but no such effect is seen in Fig. 11.4; instead the velocity curves flatten off at large radii. It seems inescapable that the radiating material of a galaxy is embedded in a halo of dark matter. Studies of the kinematics of galaxies and clusters of galaxies support the view that the density of all matter is several hundred times the density of luminous matter. If we define Ω to be the actual density divided by the critical density ρ_c, then the current value Ω_0 lies within the limits

$$0.1 \lesssim \Omega_0 \lesssim 1.0,$$

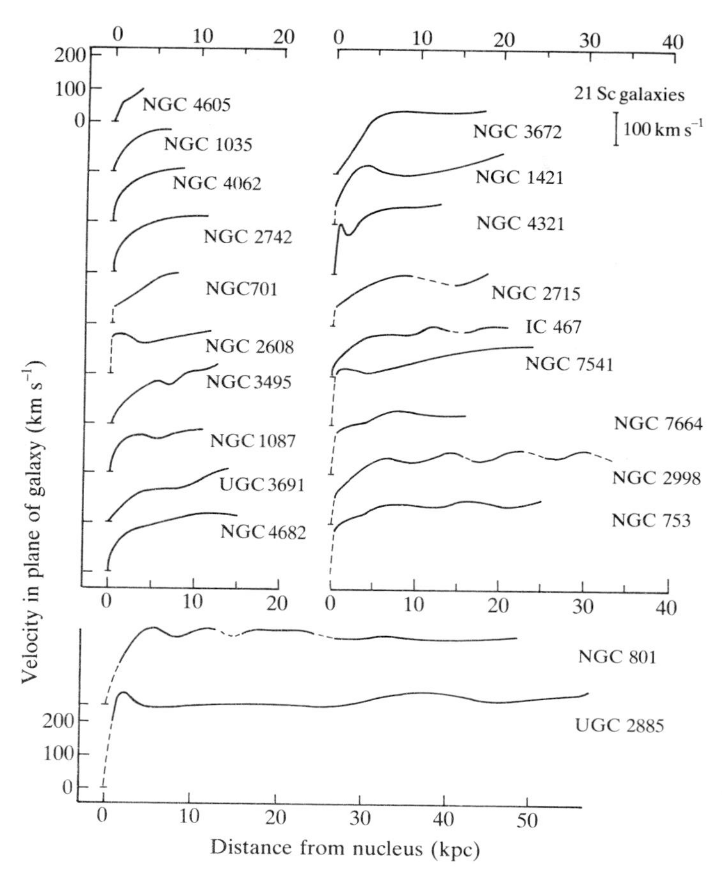

Fig. 11.4 Rotational velocities as a function of nuclear distance determined from emission line spectra for several galaxies (adapted from Rubin and Ford 1980; courtesy Professor Rubin and the *Astrophysical Journal*, published by the University of Chicago Press).

as discussed by Peebles (1986). Rewriting eqns (11.13) and (11.14) in terms of Ω, and taking Λ to be zero, gives

$$R^2 = \frac{kc^2}{H^2(\Omega - 1)}, \tag{11.17}$$

$$q = \Omega/2. \tag{11.18}$$

One simple possibility is that Ω is unity while Λ and k both vanish. Then the underlying space–time is flat, and the negative gravitational energy precisely cancels all other forms of energy. This is called an Einstein–de Sitter universe. Notice that if k is $+1$ then $\Omega_0 > 1$, while if k is -1 then $\Omega_0 < 1$.

For the present we continue to take Λ to be zero, with k unrestricted. Multiplying eqn (11.12) by 3 and subtracting eqn (11.11) gives

$$\ddot{R} = -\frac{4\pi}{3} G\rho R, \tag{11.19}$$

and so the expansion decelerates as R increases. Starting at the present epoch and extrapolating back into the past gives the solid curve shown in Fig. 11.5 with R zero at some time t_0 in the remote past, where $t_0 < H_0{}^{-1}$. As noted earlier this origin in a singularly followed by a rapid expansion is known as the 'big bang' model of the Universe. If $k > 0$ then the situation $R = 0$ corresponds to an origin at a single point.

It has been shown in the last section using eqns (11.11) and (11.12) that for the current matter-dominated universe

$$\rho R^3 = \text{constant} \equiv \rho_0 R_0^3.$$

Equation (11.11) then becomes

$$\dot{R}^2 = -kc^2 + \frac{8\pi G \rho_0 R_0^3}{3R}$$

which reduces to

$$c\,\mathrm{d}t = \frac{\mathrm{d}R}{[(1/\mu) - k]^{1/2}}$$

where $\mu = R/R_c$ and $R_c = 8\pi G\rho_0 R_0^3/3c^2$. Integrating this equation gives the following results:

$$k = +1: \qquad t = \frac{R_c}{c}[\sin^{-1}\mu^{1/2} - (\mu - \mu^2)^{1/2}] \tag{11.20}$$

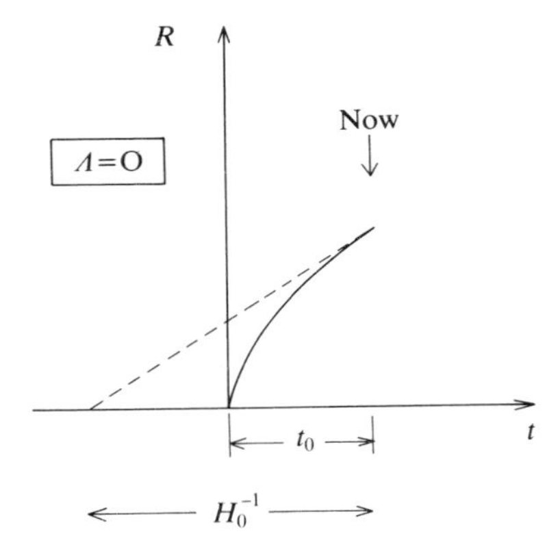

Fig. 11.5 The variation of the cosmic scale parameter R with time ($\Lambda = 0$).

$$k = 0: \qquad t = \left(\frac{2R_c}{3c}\right)\mu^{3/2} \tag{11.21}$$

$$k = -1: \qquad t = \frac{R_c}{c}[(\mu + \mu^2)^{1/2} - \sinh^{-1}\mu^{1/2}]. \tag{11.22}$$

Some examples of these solutions are exhibited in Fig. 11.6, extrapolating from conditions at present (which are marked by the cross). The relative units R/R_0 and tH_0 are employed, so that uncertainties in the values of R_0 and H_0 are masked out. When $k = +1$ and $\Omega_0 > 1$, the density of matter is large enough that ultimately its mutual attraction halts the expansion of the Universe. Thus the Universe has a maximum size and then finally collapses. For the case that $k = -1$ and $\Omega_0 < 1$ the Universe is open and expands for ever. If Ω_0 were zero the Universe would expand at velocity c following the straight line asymptote to the curves shown in Fig. 11.6. Finally if $k = 0$, and hence $\Omega_0 = 1$, the Universe is also open and unbounded but it expands ever more slowly as time passes; eqn (11.21) yields $t_0 = \frac{2}{3}H_0$. Three representative cases taken from Fig. 11.6 yield the following result:

k	Ω_0	q_0	t_0
-1	0.1	0.05	$(9\text{–}18) \times 10^9$ years
0	1.0	0.5	$(7\text{–}14) \times 10^9$ years
$+1$	2.0	1.0	$(6\text{–}12) \times 10^9$ years,

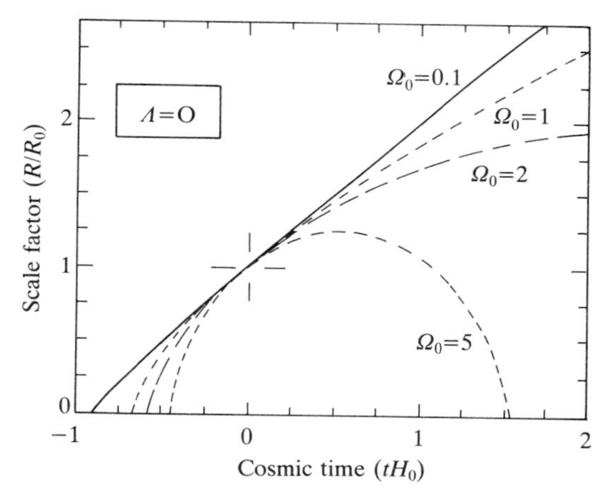

Fig. 11.6 The variation of the scale factor R/R_0 versus cosmic time tH_0 for various choices of the current density of matter ($\Lambda = 0$). A flat universe has $k = 0$ and correspondingly $\Omega_0 = 1$. A Universe with $k \leqslant 0$, i.e. $\Omega \leqslant 1.0$, will expand forever. (Adapted from Felten and Isaacman 1986; courtesy Professor Isaacman and the *Physical Review*, published by the American Physical Society.)

including the influence of the uncertainty in H_0 in estimates of the current age t_0 of the Universe. The experimental estimates discussed in Section 11.1 for the age of the Galaxy are around 15×10^9 years which seem to favour either a flat or an open Universe.

At some future time precise measurements of q should become available, and then it may be possible to infer the magnitude of the parameter R_0 which quantifies the curvature of the Universe. Taking k to be $+1$ or -1 in eqn (11.14) (and setting Λ to zero) gives

$$R^2 = \frac{c^2}{|2q - 1|H^2}$$

and

$$R_0^2 = \frac{c^2}{|2q_0 - 1|H_0^2}.$$

Experiment only restricts q_0 to the range between -1.3 and $+2.0$. Therefore only a lower limit on $|R_0|$ can be calculated, namely

$$|R_0| \gtrsim 1500 \text{ Mpc.}$$

The closer q_0 lies to 0.5 the larger the value of R_0 would be. In the limit that q_0 is exactly equal to 0.5 the Universe is flat. R_0 is a measure of the size of the Universe only if the curvature is positive. In such a case if a hypersurface through the Universe is taken at a given cosmic time t_0, R_0 is the radius of

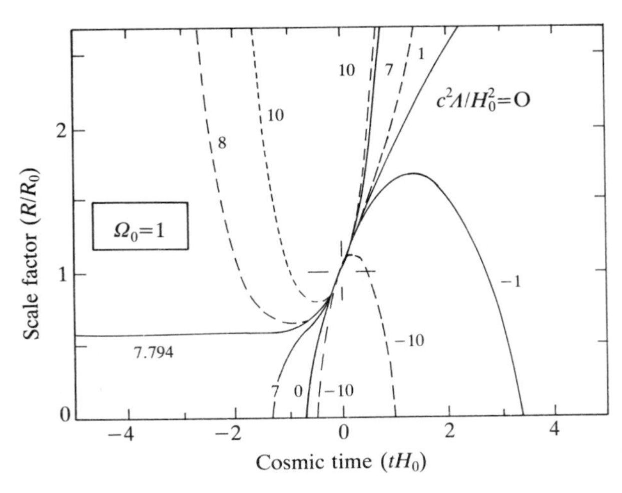

Fig. 11.7 The variation of the scale factor R/R_0 versus cosmic time tH_0 for various values of the cosmological constant ($\Omega_0 = 1$). (Adapted from Felten and Isaacman 1986; courtesy Professor Isaacman and the *Physical Review*, published by the American Physical Society.)

this hypersurface. At a distance πR_0 from the Earth on this hypersurface we would reach the antipodes of our Universe.

Finally we return to the possibility that Λ is non-zero. For simplicity we choose k to be zero, or equivalently $\Omega = \rho/\rho_c = 1$. The solutions of eqns (11.13) and (11.14) or their parent eqns (11.8) and (11.10) are shown in Fig. 11.7 for various values of Λ. One feature worthy of note is that if $|\Lambda| \approx 7H_0^2/c^2$, the lifetime t_0 of the Universe would be altered significantly.

11.5 The physical development of the early Universe

The development of the real Universe has been more subtle and interesting than the featureless expansion of an ideal fluid. There are two vital pieces of evidence for the cooling and expansion of the Universe: the CBR and the measured proportions of light nuclei (He, Li, D) in the Universe. These features are relics of phase transitions that were inevitable in a cooling expanding universe made up from matter and radiation as we know them.

First some general points need to be made. The energy density of radiation and of matter show different dependences on the scale parameter R. For matter, the particle density and hence (mass) energy density scale as R^{-3}:

$$\varepsilon_m \propto R^{-3}.$$

At present this density has the value $\varepsilon_{m0} \approx \Omega_0 \rho_c c^2 \approx 10^{-10}$ J m^{-3}, taking $\Omega_0 = 0.1$. In the case of photons their wavelength scales like R because of the cosmic red shift; the frequency and energy per photon scale as $1/R$. Therefore the energy density of cosmic background radiation is

$$\varepsilon_r \propto R^{-4}$$

The current value ε_{r0} is 4.3×10^{-14} J m^{-3}. Evidently the Universe is now matter dominated. However, the more rapid variation of ε_r with R means that at some past epoch (when $R = R_e$) there was equality between ε_r and ε_m, and at even earlier times the Universe was radiation dominated. Evaluating R_e, we obtain

$$\varepsilon_{m0}(R_0/R_e)^3 = \varepsilon_{r0}(R_0/R_e)^4$$

giving

$$R_0/R_e = 2000.$$

The cosmic background radiation would then have had a temperature of $2000 \times (2.75) = 5500$ K.

If we extrapolate back to sufficiently early times an epoch will be reached when the photon energy far exceeded the energy required to produce particle–anti-particle pairs of any particle species. In this epoch thermal equilibrium would have been maintained between all particle species with all particles moving ultra-relativistically ($v \approx c$). The total energy density would have been given by the radiation formula

$$\varepsilon = gaT^4 \tag{11.23}$$

with $g = \sum g_i/2$. Here a is the radiation constant (7.56×10^{-16} J m^{-3} K^{-4}) and g_i is a factor which takes account of the number of independent spin states. To a sufficient approximation for use here g_i is 2 for photons, electrons, and nucleons, and 1 for neutrinos. In this early Universe the term containing k in eqn (11.11) is negligible, and so the equation reduces to

$$\dot{R}^2 = 8\pi G\rho R^2/3.$$

Substituting $R_0 T_0/T$ for R and gaT^4/c^2 for ρ in this last line gives

$$-\frac{\mathrm{d}T}{T^3} = \left(\frac{8\pi Gga}{3c^2}\right)^{1/2} \mathrm{d}t.$$

Integration yields

$$T = \frac{\left(\dfrac{3c^2}{32\pi Gga}\right)^{1/4}}{t^{1/2}} = \frac{1.52 \times 10^{10}}{g^{1/4}t^{1/2}} \tag{11.24}$$

This expression can be recast in terms of the mean energy per photon 2.7 kT:

$$t \approx \frac{12}{g^{1/2}\bar{E}^2} \tag{11.25}$$

with $\bar{E}$ expressed in MeV. We also have

$$\frac{R}{R_0} = \left(\frac{32\pi Gga}{3c^2}\right)^{1/4} T_0 t^{1/2}$$
$$\approx g^{1/4} \times 1.81 \times 10^{-10} t^{1/2}. \tag{11.26}$$

In the early Universe g would have been around 50; hence $g^{1/4}$ has always lain between 1 and 2.7. Once matter came to dominate the Universe, the $t^{1/2}$ dependence of R of eqn (11.26) was replaced by the $t^{2/3}$ dependence of eqn (11.21).

As the Universe cooled so the original soup of elementary particles underwent phase transitions. Protons and neutrons would have condensed out from the plasma of quarks and gluons when kT fell to about $m_p c^2$, where m_p is the proton mass. Later these nucleons combined to form light nuclei, and still later the electrons and nuclei cooled enough for atoms to form. This latter era is known as the recombination era when matter changed from being predominantly electrons and protons to being predominantly neutral hydrogen. The transition occurred when photons energetic enough to ionize hydrogen became rare. Hydrogen has a binding energy of 13.6 eV; however, detailed calculation (Peebles 1971) taking into account the shape of the photon energy spectrum shows that the *mean* photon energy had to drop much further, to about 0.6 eV, before recombination was complete. This corresponds to a radiation temperature of about 3000 K at a time about 3×10^5 years after the big bang. Neutral hydrogen is much less effective in scattering radiation than the free protons and electrons it replaced, so that the radiation no longer remained in thermal contact with the neutral matter. The same era is therefore also called the decoupling era. Calculations show that the last time that photons in the CBR scattered from matter was at this time. The temperature at recombination agrees quite well with the estimate made above for the temperatures at the end of the radiation-dominated era. Recombination accompanied by decoupling of radiation from matter was roughly contemporaneous with the end of the radiation-dominated era.

CBR is a fossil relic from the early Universe which last interacted with matter some 10^5 years after the big bang. The fact that it is isotropic now tells us directly that the Universe was isotropic and homogeneous at the time of this last scattering. A more subtle point is that the CBR preserves a view of the Universe from an even earlier time when matter was moving relativistically with electrons and positrons in equilibrium with photons. Particle–anti-particle annihilation and bremsstrahlung at this era would have created photons while pair production would destroy them. Once the Universe cooled so that the interactions were mainly through scattering of photons from ionized matter, thermal equilibrium between matter and radiation would be maintained. This would guarantee that the photon spectrum attained or retained the *shape* of a black-body spectrum but would *not* affect the number of photons (the intensity). In fact the observed spectrum of the CBR agrees both in shape and *intensity* with a black-body spectrum, and so we can be sure that we are observing the radiation preserved from the relativistic era, long before the last scattering. Between the last scattering era and the time of the earliest quasars the matter in the Universe became inhomogeneously distributed. How this development occurred is the subject of intense current research.

The second piece of evidence from the early life of the expanding cooling Universe is the relatively large abundances of light nuclides (^{2}H, ^{3}He, ^{4}He, ^{7}Li). In round figures, out of every ten nuclei in the Universe one is helium and most of the rest are hydrogen. This proportion is independent of the sources examined (Boesgaard and Steigman 1985), which suggests that helium originated early in the Universe and not in stellar processes. In contrast the proportion of elements beyond carbon is very variable, and strongly correlated with the degree of thermonuclear processing expected for the material observed. Thermonuclear processes within stars and supernovae appear to explain well the distribution of heavy elements which constitute about 2 per cent of the visible mass of the Universe. However, these processes account for only a small fraction of the helium presently seen. In fine, an independent process is required to explain the observed helium abundance. This is equally the case for the other light elements mentioned. The processes giving most of the observed light nuclei go under the name of *primordial nucleosynthesis*.

Primordial nucleosynthesis took place during phase changes between about 1 and 300 s after the big bang in the following chronological sequence.

$t \approx 10^{-4}$ *s* Equation (11.25) shows that the particle energy $\bar{E}$ was about 100 MeV. Electrons, positrons, neutrinos, and photons were present and in thermal equilibrium. The number of nucleons present was small. For these the reactions

$$\bar{\nu}\mathrm{p} \rightleftharpoons \mathrm{e}^+\mathrm{n}$$

$$\nu\mathrm{n} \rightleftharpoons \mathrm{e}^-\mathrm{p}$$

maintained neutrons and protons at an equilibrium ratio

$$r = \frac{n}{p} = \exp\left(-\frac{\Delta mc^2}{kT}\right) \tag{11.27}$$

where Δm is the excess mass of the neutron (1.293 MeV/c^2). This ratio would have been close to unity.

$t \approx 1$ s At this time $\bar{E} \approx 1$ MeV. Cross-sections for the neutrino-initiated processes are proportional to the energy of the neutrino and would have become small at this temperature. Detailed analysis of this era when the processes converting proton to neutron became ineffective shows that the n/p ratio r froze at 0.17. This corresponds to taking $kT \approx 0.7$ MeV in eqn (11.27). A first guess as to what happened next is that the neutrons decayed with mean life about 900 s. However, matter was so dense (about 1 kg m^{-3}) that the neutrons had time to interact with protons before decaying, and the interactions gave light nuclei:

$$\mathrm{n} + \mathrm{p} \rightarrow \mathrm{d} + \gamma$$

$$\mathrm{d} + \mathrm{d} \rightarrow {}^3\mathrm{H} + \mathrm{p}$$

$$\mathrm{d} + {}^3\mathrm{H} \rightarrow {}^4\mathrm{He} + \mathrm{n}.$$

The build-up of deuterium necessary to permit the second and the third steps to occur was at first inhibited by the effect of the inverse of the first process: $\gamma + \mathrm{d} \rightarrow \mathrm{n} + \mathrm{p}$. The binding energy of deuterium is 2.225 MeV, and so once the temperature had fallen well below 1 MeV the rate of this inverse process became sufficiently slow for the helium-building processes to operate effectively.

$t > 100$ *s* By now $\bar{E} \lesssim 0.1$ MeV, and the temperature was low enough for the deuterium fraction to rise rapidly. Helium is so well bound that deuterium was quickly converted to helium: almost *all* the neutrons were effectively locked up in ^{4}He. Each ^{4}He nucleus requires two neutrons and two protons, and so it is easy to calculate the ^{4}He/(residual H) ratio from the n/p ratio r:

$$^4\mathrm{He/H} = \frac{r/2}{1-r}.$$

We take a reduced value of r (0.14) to allow for neutron decays. Then this gives a ratio ^{4}He/H close to 0.1, in excellent agreement with the observed ratio. Similar calculations for the deuterium, ^{3}He and ^{7}Li abundances also give acceptable results. There is no stable nuclide of mass 5 or mass 8, so that primordial nucleosynthesis stopped at this stage and could not build more massive nuclei. Bernstein, Brown, and Feinberg (1989) have recently presented a simplified calculation of cosmological helium production. Figure 11.8 summarizes the history of the Universe after 10^{-6} s. Particle species which were in thermal equilibrium at each epoch are entered on the figure, with the species frozen-out appended within square brackets. For black-body radiation the mean photon energy and temperature are related by the expression $\bar{E} = 2.7\,kT$.

The understanding we have of the development of the Universe permits us to put limits on the value of the cosmological constant Λ. Referring to Fig. 11.7 it is clear that values of Λ of order $+7H_0{}^2/c^2$ (10^{-52} m^{-2}) or greater are ruled out because the CBR gives us evidence of a steady expansion of the early Universe. It is equally true that Λ cannot be of order $-7H_0{}^2/c^2$ or less, because the lifetime of the Universe would then have been shorter than the observed lifetimes of the galaxies. Thus $|\Lambda| < 10^{-52}$ m^{-2}. More precise determinations of ρ_0, H_0, and q_0 are needed before we can say whether the Universe will expand forever or recollapse.

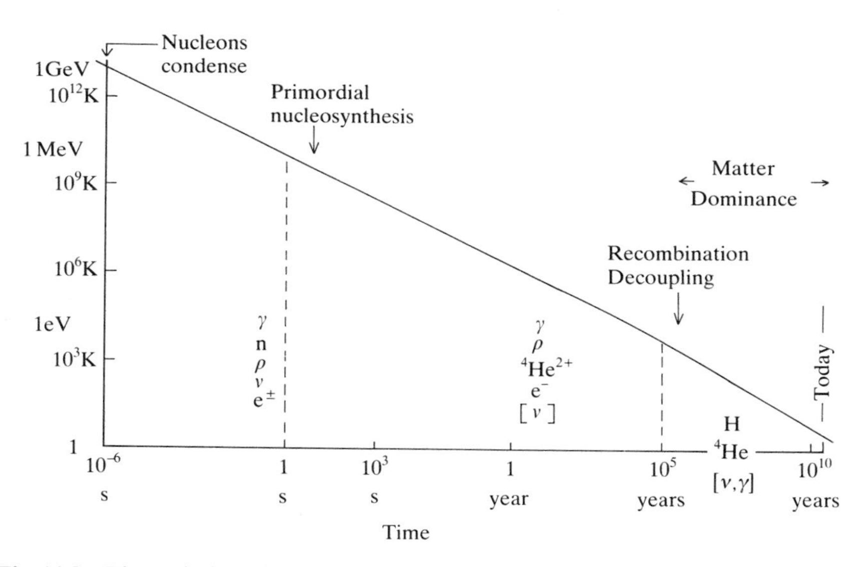

Fig. 11.8 The variation of the temperature and equilibrium particle energy versus the age of the Universe in the hot big bang model. The dominant particle species are written for several epochs. Those enclosed in square brackets are frozen out, while the other species indicated are all in thermal contact.

11.6. Olbers' paradox

The fact that the sky is dark was mentioned in Section 11.1 as an indication that the Universe cannot be unchanging and of infinite extent. Assume for the moment that this were the case. Then for a region of the sky where the average distance to the nearest galaxies is r it follows that the number of galaxies visible is proportional to r^2 while the apparent brightness of each is proportional to $1/r^2$. Hence the sky should be uniformly bright! This paradox is named after Heinrich Olbers who noted it in 1826. The paradox is resolved by noting that the galaxies have only existed for a finite length of time, so that in some regions of the sky no galaxies are in view. This argument can be made quantitative as follows.

Let us calculate how much light reaches us from successive shells around the Earth and sum over all shells containing galaxies. A shell of radius σ and thickness $d\sigma$ in comoving coordinates has a volume

$$\begin{aligned} dV &= \int_{\theta=0}^{\pi} \int_{\varphi=0}^{2\pi} \frac{R\, d\sigma}{(1-k\sigma^2)^{1/2}} (R\sigma\, d\theta)(R\sigma \sin\theta\, d\varphi) \\ &= \frac{4\pi R^3 \sigma^2\, d\sigma}{(1-k\sigma^2)^{1/2}}. \end{aligned}$$

The density of galaxies in this shell is given in terms of the current density of galaxies n_0 by

$$n = n_0 \left(\frac{R_0}{R}\right)^3.$$

Suppose that the luminosity of a galaxy is L_0 W. The gravitational red shift reduces the energy of each photon by a factor R/R_0, and the interval between photons increases by this same factor R/R_0. As a result the apparent luminosity is reduced to $L_0(R/R_0)^2$, where R and R_0 are the values of the curvature parameter when the radiation is emitted and when it is received on Earth respectively. The energy flux per unit area on Earth from one galaxy in the shell is thus

$$F = \frac{L_0(R/R_0)^2}{4\pi R_0{}^2 \sigma^2}.$$

The total energy flux received from all galaxies in the shell is

$$nF\, dV = \frac{n_0 L_0 R^2\, d\sigma}{R_0(1-k\sigma^2)^{1/2}}.$$

Along the path of a photon to Earth the metric equation (11.3) simplifies to

$$c\, dt = \frac{R\, d\sigma}{(1-k\sigma^2)^{1/2}}.$$

Hence we can write

$$nF\,\mathrm{d}V = n_0 L_0 (R/R_0) c\,\mathrm{d}t.$$

Integrating this expression from the time t_f at which galaxies first formed up to the present time t_0, gives the total luminious flux on Earth:

$$Q = n_0 L_0 c \int_{t_f}^{t_0} (R/R_0)\,\mathrm{d}t.$$

Ignoring the expansion of the Universe we would obtain

$$Q_s = n_0 L_0 (t_0 - t_f) c.$$

Now $n_0 L_0 = 0.4 \times 10^{-32}$ W m^{-3} and $t_0 - t_f \approx 1.5 \times 10^{10}$ years so that

$$Q_s \approx 0.5\ \mu\mathrm{W\ m}^{-2}.$$

This value is extremely small, about one millionth of the illumination of a book 2 m from a 100 W lamp. Already we can understand why the night sky is dark. Taking into account the expansion of the Universe makes only a small changes: Wesson, Valle, and Stabell (1987) show that, for any reasonable model,

$$Q \approx 0.5 Q_s.$$

This makes it quite clear that the sky is dark because the galaxies have existed for only a finite length of time. The energy density of the radiation from galaxies is around Q/c, i.e. 10^{-15} J m^{-3}, smaller in fact than the energy density of the microwave background (4.3×10^{-14} J m^{-3}).

11.7 Inflation?

The outline of the early history of our Universe given in Section 11.5 is generally accepted. However, it does suffer from difficulties, partly philosophical, of which the horizon problem is the most troublesome. At any given event in space–time we can define the particle horizon, marked out by the most distant events from which light could have been received during the past life of the Universe. We say that an event is causally connected to events within its horizon. Now because the CBR seen from Earth is isotropic to a high degree it might seem reasonable to expect that the whole of the region we can see (the visible Universe) is causally connected; otherwise how did it reach the same temperature to better than one part in 10^4? However, this simple inference is invalid! Consider two events α rad apart from which we now receive CBR. Their separation is approximately $\alpha c t_0$ where t_0 is the present age of the universe (about 1.5×10^{10} years). During the matter-dominated era the growth rate has been like $R \propto t^{2/3}$. Therefore in the decoupling era (3×10^5 years) the universe was 1200 times smaller than at present, and the two regions of CBR were only a distance $\alpha c t_0/1200$ apart. Taking α to be 2^0, we find that at decoupling the separation was 4×10^{21} m. In comparison light travelling since the origin of the Universe, 3×10^5 years earlier, would only have travelled about 3×10^{21} m. Evidently the two regions could not have been in causal contact! The situation is aggravated when we consider regions of the sky which are 180^0 apart.

One way of resolving this problem is to imagine that all parts of the Universe were initially identical although never in causal contact and that they would have developed in the same manner. An objection to this argument is that at very early times the Universe was dominated by quantum fluctuations, which would have produced uncorrelated effects at different locations.

A more consistent explanation for the uniformity of the *visible* Universe is possible if the whole Universe underwent a very rapid expansion early in its existence ($t \ll 10^{-6}$ s). In this way a small causally connected volume could have been expanded to the size of the visible Universe. This sudden expansion also provides an explanation of why the Universe is now close to being flat. Figure 11.9 shows how the horizon and size of the Universe now visible would behave without inflation, where the expansion is taken to follow the relation $R \propto t^{1/2}$ appropriate to a radiation-dominated universe. Figure 11.10 shows the effect of inflation by a factor of 10^{50} occurring somewhere around 10^{-36} s. In this inflationary period the fabric of space–time is required to stretch at a rate many orders greater than the speed of light. Such a dramatic expansion would lead to a horizon distance orders of magnitude larger than the volume

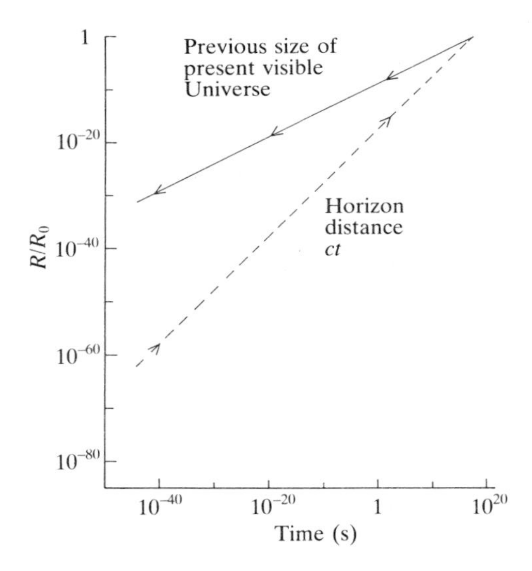

Fig. 11.9 The broken line indicates how the distance to the horizon increased through the life of the Universe. The solid line extrapolates the size of that part of the Universe which we now see back into the past. This diagram is appropriate for the standard big bang model.

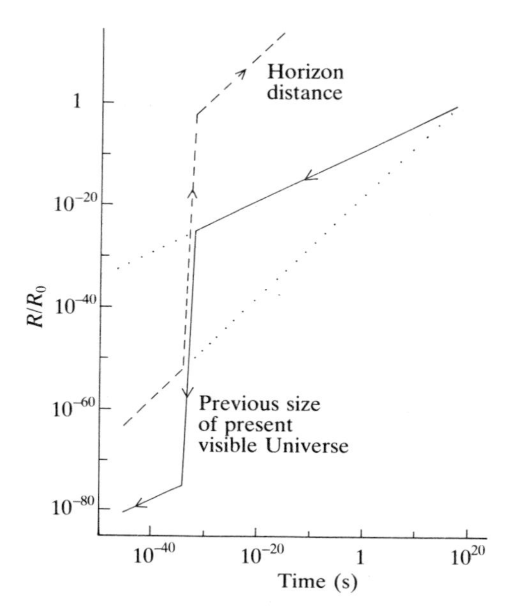

Fig. 11.10 This diagram shows how the development of Fig. 11.8 is modified when there is inflation early in the life of the Universe.

of the Universe we now observe, thus solving the horizon problem. After inflation the Universe would inevitably be nearly flat, as indeed it is.

Many mechanisms have been proposed for inflation (see Guth and Steinhardt 1984, Linde 1987), but only one will be discussed here. This is the process of spontaneous symmetry breaking, a type of phase transition likely to occur in the early Universe. In addition to the familiar spin 1 (photons) and spin $\frac{1}{2}(e^{\pm})$ particles, the theories of elementary particles predict the existence of the Higgs particles. These would have the quantum numbers of the vacuum (spin zero) so that a phase change is possible in which potential energy from the Higgs field is dumped into the vacuum, a process which could not happen in the case of the electromagnetic (photon) field. This is equivalent to making the cosmological constant temporarily large and positive. It is thought likely that one such phase transition would occur when the strong and electroweak forces reach comparable strengths, i.e. at an energy per particle of about 10^{15} GeV, which would also set a time of about 10^{-36} s and a Higgs mass of about 10^{15} GeV/c^2. The effect of increasing the cosmological constant drastically during this phase transition is to reduce eqn (11.8) to the form

$$3\dot{R}^2/R^2 = c^2\Lambda_i,$$

with a solution

$$R \propto \exp[ct_i(\Lambda_i/3)^{1/2}]$$

where t_i is the length of the inflation period. Subsequent to the expansion the Universe is quite flat and virtually empty. The Higgs particles then decay to give the quark, lepton, and gluon plasma. The successive later stages in the life of the Universe follow the pattern we have already discussed. Whether inflation took place and the details of the mechanism are still matters of speculation.

12
The path to quantum gravity

By the year 1974 quantum field theories had been developed to describe the weak, electromagnetic, and strong forces, but there is still no accepted quantum theory for the remaining force, gravity. The underlying difference between gravitation and the other forces is that in the case of gravitation it is the structure of space–time itself that must be quantized. In Section 12.1 the Planck scale is introduced, namely the scale at which quantum and general relativistic effects are both important. Section 12.2 is used to outline the techniques which have been successful in building the existing quantum field theories. A direct application of this approach encounters basic difficulties in the case of gravitation. Some remarks are made about the equivalence principle and tests using elementary particles in Section 12.3. In section 12.4 possibilities for the unification of forces are discussed. One thread in the discussion concerns the discovery by Kaluza that it is possible to unify gravitation and electromagnetism at the classical level. At present the strongest contender for a quantum field theory to describe all forces is the supersymmetric string theory. This development will be discussed in Section 12.5.

12.1 The Planck scale

A useful plot for comparing regions where general relativistic and quantum effects become important is Figure 12.1 which shows size versus mass. General relativistic effects become dominant in space–time when a star collapses to within its Schwarzschild radius $2GM/c^2$. The upper sloping line $r = 2GM/c^2$ on Fig. 12.1 marks this transition: anything to its left is a black hole. Quantum fluctuations are similarly important to the left of the lower sloping line which has the equation $R = \hbar/Mc$ giving the Compton wavelength of a mass M. In order to understand this assertion we consider a fluctuation of energy E in which a particle–anti-particle pair are created from the vacuum and subsequently annihilate. The duration of the fluctuation cannot exceed a time

$$t = \hbar/E$$

given by the uncertainty principle, and so the range of influence of the fluctuation is less than

$$R = ct = c\hbar/E.$$

Thus for a mass M, the uncertainty in position due to quantum fluctuations

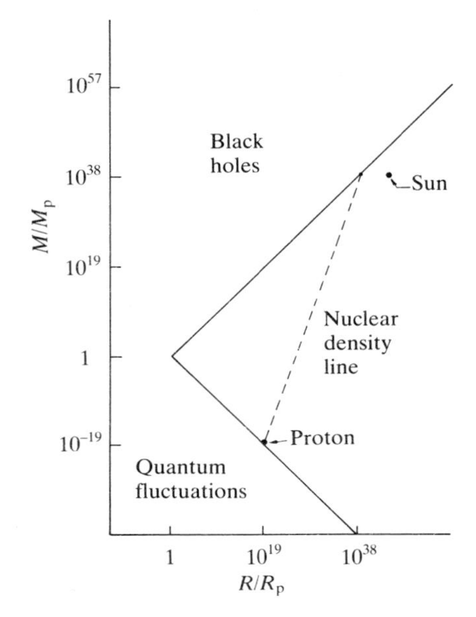

Fig. 12.1 Mass–radius diagram showing the regions where general relativistic and quantum effects are important. M_p is the Planck mass and R_p the Planck length.

is of order $R = \hbar/Mc$. At sufficiently small distances and large enough masses both general relativistic and quantum effects must become important. This condition is reached where the two lines intersect on Fig. 12.1. Then (dropping the factor 2, as is customary)

$$GM_p/c^2 = \hbar/M_p c$$

giving a mass

$$M_p = (\hbar c/G)^{1/2} = 2.18 \times 10^{-8} \text{ kg}$$

and a distance

$$R_p = (\hbar G/c^3)^{1/2} = 1.62 \times 10^{-35} \text{ m}.$$

R_p is known as the Planck distance because Planck was the first to notice that this combination of the fundamental constants $\hbar$, c, and G gave a natural scale of length. The corresponding Planck energy measured in GeV ($=10^9$ eV) is

$$E_p = M_p c^2 = 1.22 \times 10^{19} \text{ GeV},$$

and quantum fluctuations of this size can persist for a period called the Planck time:

$$t_p = \hbar/M_p c^2 = 5.31 \times 10^{-44} \text{ s}.$$

On the Planck scale therefore a quantum fluctuation can generate a black hole, and in order for there to be overall consistency between physical theories it seems inescapable that GR (gravitation) should be quantized. There is already an implication that this is necessary inherent in the form of the Einstein equation

$$G_{\mu\nu} = \frac{8\pi G}{c^4} T_{\mu\nu}.$$

$T_{\mu\nu}$ is constructed from four momenta which are quantized, so that it is unnatural to leave $G_{\mu\nu}$ as a classical quantity. The waves associated with gravitation are oscillations of space–time which means that the structure of space–time itself is to be quantized. With the quantization of space–time the concept of a smooth space–time continuum is lost, and this leads to fundamental conceptual difficulties which we shall explore later. In the case of electromagnetism the quanta of the vector field A_μ are vector (spin 1 in units of $\hbar$) particles, namely photons. Correspondingly, with gravitation, whose field $G_{\mu\nu}$ is a rank 2 tensor, the quanta would be tensor (spin 2) particles known as gravitons.

Quantities on the Planck scale are remote from our experience; none the less if the history of the Universe is traced back towards the big bang, then an epoch is eventually arrived at when the energy per particle exceeded the Planck energy. This is called the *Planck era.* Let us take the simple case of a flat radiation-dominated Universe: time and temperature are then related by eqn (11.24). We write this with g set to unity:

$$t = \frac{(3c^2/32\pi Ga)^{1/2}}{T^2}$$

Re-expressing this in terms of the energy per particle $\bar{E} = kT$ and writing the radiation constant in full as $a = \pi^2 k^4/15c^3\hbar^3$ gives

$$t = \frac{(45\hbar^3 c^5/32G\pi^3)^{1/2}}{\bar{E}^2}.$$

The mean energy would therefore have exceeded the Planck energy at times earlier than

$$\frac{(45\hbar^3 c^5/32G\pi^3)^{1/2}}{E_\mathrm{p}^2} = \left(\frac{45}{32\pi^3}\right)^{1/2} t_\mathrm{p}.$$

Thus the Planck era endured for a time of order t_p after the big bang. It follows that a complete description of the evolution of the very early Universe must be based on a quantum theory of gravity. Restricted calculations have been made using classical space–time with quantum fluctuations that preserve the causal structure, with or without quantum fields for other particle species. These semi classical calculations suggest that an initial singularity at the origin of the Universe can be avoided.

12.2 Quantum gauge theories

In the current epoch the gravitational and electromagnetic forces have infinite range, while the other fundamental forces have short ranges. The strong force responsible for binding quarks into nucleons, and nucleons into nuclei, has a range of 10^{-15} m; the weak force which contributes to energy-generating processes inside the Sun has an even shorter range of 10^{-17} m. There is an equally striking contrast between the strengths of these forces. For two protons in contact they are in the ratio

$$\text{strong} : \text{e.m.} : \text{weak} : \text{gravity} = 1 : 10^{-2} : 10^{-7} : 10^{-38}.$$

It is a fact of major significance that the theories of *all* forces are based on what is called the *gauge principle.* This states that physical theories should be independent of the local choice of coordinates, where these can change in any arbitrary but smooth manner across space–time. In GR the gauge principle takes the form of the principle of generalized covariance discussed in Chapter 6. When the formulae of SR are rewritten with covariant derivatives replacing normal space–time derivatives these formulae become valid under general coordinate transformations. The gauge theories of the strong, electromagnetic, and weak forces differ structurally from GR because they involve *internal* spaces not directly accessible to our senses. The case of electromagnetism is the simplest and will therefore be the one we discuss here; its quantum field theory is known as quantum electrodynamics (QED).

In elementary quantum mechanics an electron is described by a wavefunction $\psi(\boldsymbol{r}, t)$ which is a function of position and time. The momentum of the electron is the expectation value of the operator $-\mathrm{i}\hbar\partial/\partial x^{\mu}$:

$$p_{\mu} = -\mathrm{i}\hbar \int \psi^* \left(\frac{\partial \psi}{\partial x^{\mu}}\right) \mathrm{d(volume)}.$$

The probability of finding the electron in a volume element is

$$\psi^*\psi \ \mathrm{d(volume)}.$$

The overall phase of ψ has no significance. If ψ is changed to $\psi \exp(\mathrm{i}\varphi e)$, where e is the charge and φ a constant, then neither $\psi^*\psi$ nor $\psi^*\partial\psi/\partial x^{\mu}$ is altered. This type of transformation is called a global gauge transformation. Of more interest are the local gauge transformations analogous to the transformations of GR, where φ is dependent on position in space–time $\varphi(\boldsymbol{r}, t)$. Things then change markedly:

$$\psi \rightarrow \psi \exp[\mathrm{i}e\varphi(\boldsymbol{r}, t)] \tag{12.1}$$

so that the electron momentum integrand

$$\psi^* \frac{\partial \psi}{\partial x^{\mu}} \rightarrow \psi^* \frac{\partial \psi}{\partial x^{\mu}} + \mathrm{i}e|\psi|^2 \frac{\partial \varphi}{\partial x^{\mu}},$$

which is clearly different from before. This sort of effect has been met previously in developing GR, and the answer there was to define a covariant derivative that automatically compensates for any generalized coordinate transformation. There is a crucial distinction that here the transformation (12.1) acts not on space–time but on the wavefunction. Changes in φ cause a rotation of ψ in an Argand diagram which is called charge space because the effect depends on the charge. Charge space is an *internal* space not connected with space–time. The gauge transformations described by eqn (12.1) form a group which is called U(1) and the electromagnetic force is said to possess a U(1) symmetry. In the name U(1) the U specifies that the transformations of the group are unitary (leaving the lengths and angles between the ψ vectors in charge space unchanged) and the 1 indicates the single complex dimension of charge space.

The covariant derivative needed to ensure that the electron momentum is coordinate independent has the form

$$\frac{\mathrm{D}}{\mathrm{D}x^{\mu}} = \frac{\partial}{\partial x^{\mu}} + \mathrm{i}eA_{\mu} \tag{12.2}$$

and in consequence the momentum definition is altered to

$$p_{\mu} + \mathrm{i}eA_{\mu} = -\mathrm{i}\hbar \int \left(\psi^{*}\frac{\partial\psi}{\partial x^{\mu}} + \mathrm{i}e\psi^{*}A_{\mu}\psi\right)\mathrm{d(volume)}. \tag{12.3}$$

A_{μ} is a vector which does the duty of the metric connection in GR. When this differential is used

$$\psi^{*}\frac{\mathrm{D}\psi}{\mathrm{D}x^{\mu}} \rightarrow \psi^{*}\frac{\mathrm{D}\psi}{\mathrm{D}x^{\mu}}$$

under the transformation of eqn (12.1) provided that

$$A_{\mu} \rightarrow A_{\mu} - \frac{\partial\varphi}{\partial x^{\mu}}. \tag{12.4}$$

Equations (12.3) and (12.4) are well-known results in classical electromagnetism with A_{μ} being identified as the four-vector electromagnetic field. In that theory eqn (12.4) defines local gauge transformations under which the measurable quantities, such as the electric and magnetic field, do not alter. $p_{\mu} + \mathrm{i}eA_{\mu}$ is then the four-momentum of the electron plus the electromagnetic field, and is a conserved quantity. Evidently the U(1) symmetry we have described was already latent in the classical description of electromagnetism.

Calculations in QED, the quantum field theory of electromagnetism, make use of what are known as Feynman diagrams. The term $\psi^{*}A_{\mu}\psi$ in eqn (12.3) that describes how the electron interacts with the electromagnetic field takes on a new interpretation. ψ specifies the annihilation of an incident electron, A_{μ} the emission of a photon, and ψ^{*} the creation of another electron, all at the same point in space–time. This process is shown in Fig. 12.2 and is called

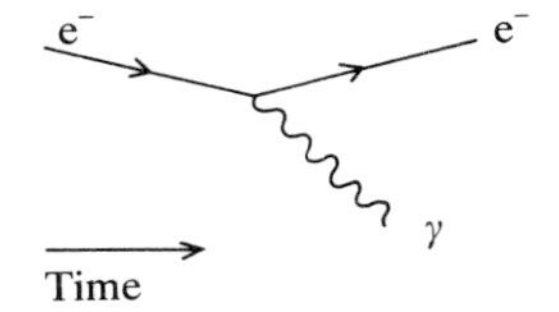

Fig. 12.2 The vertex for electromagnetic interactions.

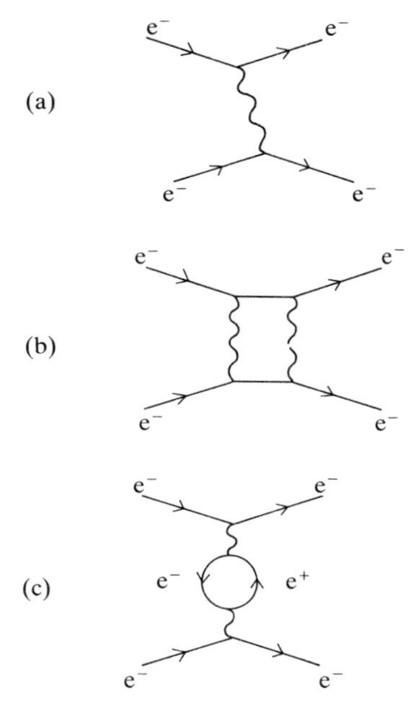

Fig. 12.3 Feynman diagrams for electron–electron scattering with (a) single-photon exchange, (b) two-photon exchange, and (c) single-photon exchange with one internal loop.

a vertex. Figure 12.3(a) shows how vertices can be joined to give the simplest Feynman diagram in which two electrons scatter by exchanging a single photon. The amplitude for the process contains a factor e from each vertex so the amplitude is proportional to e^2 or more precisely to

$$\alpha = \frac{e^2}{4\pi\varepsilon_0 \hbar c} \approx \frac{1}{137},$$

the fine structure constant. The rate of any process is proportional to the amplitude squared, and so here the rate is proportional to α^2. Multiple photon exchanges can occur and the total amplitude is the coherent sum of the amplitudes for all possible diagrams. Each additional internal line as in

Fig. 12.3(b) imparts a new factor α^2 in the rate, so that diagrams with the fewest exchanges are the most important. Very precise predictions can be made using only a limited number of diagrams just because α is so small. What is more, these predictions are in excellent agreement with *all* experimental measurements of pure electromagnetic effects.

Figure 12.3(c) illustrates a diagram with an internal loop involving the creation and subsequent annihilation of an electron–positron pair. Special techniques must be used to handle these diagrams; these techniques work for QED but fail when the attempt is made to quantize gravity. If the electron–positron pair is short-lived, then according to the uncertainty principle, they can have very large energy. In this limit the creation and annihilation vertices coalesce, and the amplitude for the diagram becomes infinite! However, all such infinite contributions can be absorbed by introducing two infinite constants into the theory connected to mass and charge. For example, the observed mass is given by

$$m = m_0 - m'$$

where m_0 is the *bare* mass of a non-interacting electron and m' is the correction due to its interaction with photons. Both m_0 and m' are infinite, but m retains its finite measured value. Theories in which the infinite contributions can be absorbed by a finite number of (unobservable) infinite constants are known as *renormalizable*.

Unfortunately, when the attempt is made to construct Feynman diagrams and rules for gravity with gravitons replacing photons the resulting theory is not renormalizable. This comes about because the fine structure constant α, which is dimensionless, is replaced by G, which has dimensions. Then each diagram with additional internal loops has additional powers of G in its amplitude; in turn, the infinite factor in a loop amplitude has different dimensionality from the infinite factor arising in simpler diagrams. The result is a totally unmanageable theory.

12.3 The equivalence principle

It was concluded in Chapter 2 on the basis of very precise experiments that the ratio of the gravitational to the inertial mass is the same for all matter. What implications does this have at the level of the elementary particle? In matter each electron is continuously exchanging photons with the nucleus to which it is bound and a proportion of these photons convert to electron–positron pairs. Exchanges also occur between the nucleons, with mesons being the exchanged particles. Hence the equality of gravitational and inertial mass requires that the mass attributable to the exchanged particle feels the same gravitational force as any other form of matter.

Direct measurements on elementary particles in free fall are hard to carry out, and they yield nowhere near the precision attained with macroscopic bodies. Koester (1976) has compared the gravitational and inertial masses of neutrons and finds a ratio of 1.00016 ± 0.00025. When the rate of free fall of a charged particle, such as an electron, is to be measured there is the additional complication that very weak electric fields can easily mask the gravitational force. The electric field which will exactly match the gravitational force on an electron is

$$E = mg/e \approx 5 10^{-11} \text{ V m}.$$

Therefore the experiment needs to be performed in a region free from electric fields, and we might guess that a Faraday cage in the form of a long vertical metal cylinder would provide the necessary shielding. However, the conduction electrons and the lattice ions in a metal are also acted on by gravity, and their movement leads to the establishment of electric fields in the metal. Considering the conduction electrons alone, their concentration near the bottom of the metal gives rise to a downward electric field that tends to push electrons upward. In equilibrium the field is mg/e. Lattice ion movement produces an opposite field which is expected to be larger. Witteborn and Fairbank (1968) and Lockhart, Witteborn and Fairbank (1977) have studied these complications. Finally it is worth remarking that it may be feasible to measure the acceleration of anti-protons in free fall now that techniques to store large numbers (10^{11}) of anti-protons have been mastered at the European Centre for Particle Physics (CERN). Electric shielding is less critical for anti-protons than for electrons because the anti-proton mass is a factor of 2000 larger.

12.4 The unification of forces

The weak and strong forces are described by renormalizable theories constructed along the lines indicated in Section 12.2. Each is based on an underlying symmetry in an internal space—a two-dimensional space for the weak force with a symmetry group SU(2) and a three-dimensional space for the strong force with a symmetry group SU(3). It is a notable feature of quantum gauge theories that the strength of the force changes as the total centre-of-mass energy increases; this is illustrated in Fig. 12.4. The solid lines show the experimentally observed variation, while the broken lines are inferences from the theory. As the energy rises to about 300 GeV the weak electromagnetic forces converge to a single electroweak force, while over the same range the strong force declines in strength. If the predicted variation above this energy can be relied on, then at some energy around 10^{15} GeV the strong and electroweak forces should unify: this state is called grand unification.

The energy per particle in the Universe during the early expansion would have proceeded across Fig. 12.4 from right to left. What happens at each change of phase when the forces bifurcate remains obscure in detail. One new feature is that the state of the vacuum is important. In classical mechanics

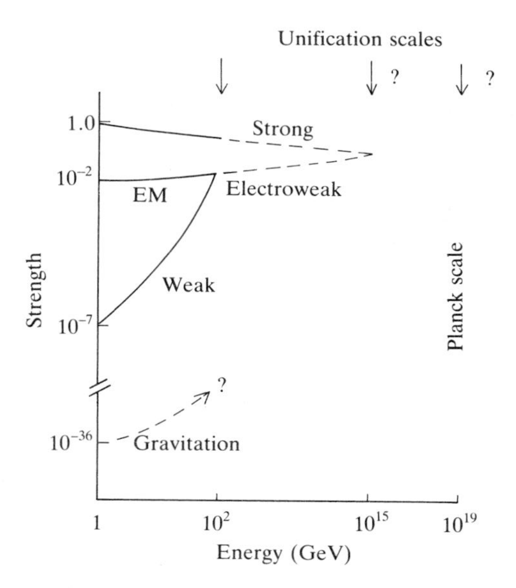

Fig. 12.4 The measured (solid lines) and possible (broken lines) variation of the interaction strengths as a function of centre-of-mass energy in two-particle interactions.

the vacuum is a unique empty state; in contrast, the quantum-mechanical vacuum is in continuous activity, with spontaneous creation and annihilation of particle–anti-particle pairs occurring anywhere. The process of unification in its simplest form brings in scalar particles (the Higgs bosons) and it is then possible to have alternative vacua because scalar particles have the same quantum numbers as the vacuum. A new vacuum state could be built from the standard quantum vacuum by adding a uniform scalar field density ρ everywhere. Above the unification temperature T_u the vacuum with lowest energy could have one value ρ_+ of the scalar field density, while below the unification energy a vacuum with a different constant scalar field density ρ_- would be the vacuum of lowest energy. During the cooling of the Universe the transition from a scalar field density ρ_+ to ρ_- occurs at T_u. An enormous energy release is possible which would, for suitable scalar field parameters, lead to the inflation of the Universe discussed in Chapter 11. The sort of transition envisaged for the vacuum at the unification temperature T_u is quite familiar in solid state physics: energy is released at a phase transition when one atomic arrangement replaces another as the lowest energy configuration. The grand unification energy scale (about 10^{15} GeV) is not far removed from the Planck scale (about 10^{19} GeV) at which all forces may very well be unified and of equal strength.

Another pathway in the direction of unifying the forces of nature led directly from GR. The first steps along this path were taken by Kaluza (1921) and

Klein (1926). Kaluza, then a teacher at Königsberg, discovered that if GR is written for a five-dimensional space–time it neatly breaks down into standard GR plus electromagnetism, with the fifth (space) dimension curled up into a circle of radius similar to the Planck distance. This is a very attractive result because this compact dimension matches the internal 'charge' dimension postulated in the gauge theory of electromagnetism. In each case there can be rotations around the origin but no displacement. One ingredient in Kaluza's argument is the electromagnetic tensor

$$F_{\mu\nu} = A_{\mu,\nu} - A_{\nu,\mu} \tag{12.5}$$

in terms of which the inhomogeneous Maxwell equations can be written

$$F_{\mu\nu,}{}^{\mu} = j_\nu \tag{12.6}$$

where j_μ is the four-vector current. Another ingredient is to write GR for four spatial dimensions and one time dimension. A component of the metric connection involving the fifth dimension is, using eqn (6.5),

$$2\Gamma_{\mu 5\nu} = g_{5\mu,\nu} - g_{5\nu,\mu} + g_{\mu\nu,5}.$$

Now if the fifth dimension is curled up as in 'charge' space, the derivatives with respect to this dimension vanish: $g_{\mu\nu,5} = 0$. Consequently the last equation reduces to

$$2\Gamma_{\mu 5\nu} = g_{5\mu,\nu} - g_{5\nu,\mu}. \tag{12.7}$$

Turning now to the Einstein equation, the terms involving the fifth component *in vacuo* are

$$R_{5\nu} = \frac{8\pi G}{c^4} T_{5\nu}.$$

Specializing to the frame in free fall gives

$$R_{5\nu} = \Gamma^{\lambda}_{5\nu,\lambda} - \Gamma^{\lambda}_{5\lambda,\nu}.$$

However, using the symmetry of $g_{\lambda\beta}$, we obtain

$$\Gamma^{\lambda}_{5\lambda} = g^{\lambda\beta} g_{\lambda\beta,5}/2,$$

which is zero because the fifth dimension vanishes. Thus

$$\begin{aligned} R_{5\nu} &= \Gamma^{\lambda}_{5\nu,\lambda} \\ &= g^{\lambda\beta}\Gamma_{\beta 5\nu,\lambda}, \end{aligned}$$

and Einstein's equation becomes

$$g^{\lambda\beta}\Gamma_{\beta 5\nu,\lambda} = \frac{8\pi G}{c^4} T_{5\nu}. \tag{12.8}$$

Kaluza appreciated that it is possible to make correspondences between the metric components and electromagnetic field components

$$g_{5\mu} \equiv A_{\mu} \tag{12.9}$$

and have a consistent theory. In the four normal dimensions standard GR is reproduced, while for the fifth dimension we can now show that the theory of electromagnetism is recovered. Using the ansatz (12.9) in eqn (12.7) gives

$$\Gamma_{\mu 5 \nu} = F_{\mu\nu}.$$

In turn, eqn (12.8) becomes

$$F_{\beta\nu,}{}^{\beta} = \left(\frac{8\pi G}{c^4}\right) T_{5\nu}$$

which reproduces the inhomogeneous Maxwell eqn (12.6) if

$$j_{\nu} \equiv \left(\frac{8\pi G}{c^4}\right) T_{5\nu}.$$

The homogeneous Maxwell equations

$$F_{\mu\nu,\lambda} + F_{\nu\lambda,\mu} + F_{\lambda\mu,\nu} = 0$$

follow automatically when eqn (12.7) is used. This is an altogether remarkable result, especially as the compact fifth dimension can be identified with the complex dimension necessary for the quantum gauge theory QED. This dimension can be viewed as curled up into a cylinder so that only the rotation angle ($e\varphi$ in the notation of Section 12.2) is a free variable. Further analysis shows that the radius of the cylinder is of the order of the Planck distance.

If the Kaluza prescription is generally valid then the following statement can be made: gauge fields of forces are the components of the metric in some higher-dimensional space–time, and these extra spatial dimensions are compact with size of order of the Planck distance. Another equally unproven but attractive possibility is that while all spatial dimensions were compact in the Planck era, the three dimensions of normal space subsequently underwent inflation during the phase transitions occurring in the early life of the Universe. The other spatial dimensions remain compact to this day.

12.5 Supersymmetric string theory

At energies above the scale for unifying the weak and electromagnetic forces these two forces are subsumed to a single electroweak force with the compound symmetry $SU(2) \otimes U(1)$. Above the grand unification scale it seems likely that the strong and electroweak forces blend into a single force with a larger symmetry. This trend to greater symmetry in the early Universe makes it of interest to investigate other symmetries that could come into play. One which encompasses fermions (with half-integral spin and obeying Fermi–Dirac statistics) and bosons (with integral spin and obeying Bose–Einstein statistics) is called supersymmetry. This symmetry would have the effect of unifying matter and radiation, one of the major dichotomies of the present Universe. Supersymmetry is in fact the only remaining space–time symmetry that is compatible with Lorentz invariance: the others are symmetry under displacements in space–time, rotations in space, and Lorentz boosts. Supersymmetry would require invariance under the transformation

$$\text{fermion} = Q \text{ boson}$$

where Q is called a spinorial charge. According to the rules of group theory the product of the Qs is also a valid transformation; explicitly

$$Q_iQ_j + Q_jQ_i = C_{ijk}P_k$$

where C_{ijk} is a constant and P_k is the four-momentum. Note that the stress-energy tensor is constructed from four-momenta, and so we can see a link here with GR.

Supersymmetric theories require that every known boson (fermion) should have a supersymmetric partner which is a fermion (boson). Thus the electron should have a partner, the selectron, which is a scalar (spinless) particle. The supersymmetric particles would interact with the usual gauge coupling strengths, and the interactions of the lightest and stable supersymmetric particle (probably the photino) with normal matter is expected to be weak. Gravitational attraction between all particles, whether supersymmetric or not, would have its normal strength. Direct evidence for the existence of supersymmetric particles would be hard to come by, and so far none have been

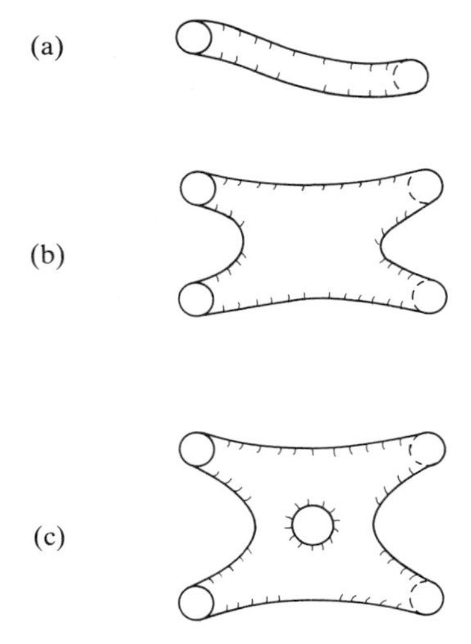

Fig. 12.5 Diagrams showing the world sheets traced out by closed strings in space–time; (a) a single string; (b) two strings that scatter; (c) two strings that scatter and have an internal loop.

observed. One possibility mooted is that some of the dark matter in the Universe might be supersymmetric. Schemes with supersymmetry provide a way of avoiding the unification of electromagnetic, weak, and strong forces at the same energy scale.

Attempts have been made using supersymmetry to construct gauge theories which include all forces, and which explicitly quantize the gravitational field. However, the supersymmetric and Kaluza–Klein-based schemes for unifying gravity with the other forces are likely to come to grief for the same reason: they are all non-renormalizable. Nevertheless these attempts have revealed basic connections between the fundamental forces which must certainly be present in any successful theory. These connections have been the ones stressed above.

One model that is expected to overcome the problems of quantizing gravity is superstring theory (Schwarz 1984, 1985). The 'super' in superstrings indicates that supersymmetry is a component of the theory; 'string' is the vital new feature. Particles, which have hereofore been thought of as points in space–time, are now to be pictured instead as one-dimensional strings of approximately the Planck length 10^{-35} m. A particle therefore traces out a two-dimensional world-sheet rather than a world-line as it travels through space–time. This is shown for a string in the form of a closed loop in Fig. 12.5(a). Figure 12.5(b) shows a particle–particle interaction and Fig. 12.5(c) a loop diagram. Figure 12.5(b) is the analogue of Fig. 12.3(a) and Fig. 12.5(c) is the analogue of Fig. 12.3(b) as well as of 12.3(c). The effect of spreading the particle out along a line of length R means that the momentum uncertainty of the particle is

$$P \approx \hbar/R.$$

As a result the integrals over kinematic variables stop at $\hbar/R$ whereas they extended to infinite momenta for point particles. The amplitude for diagram 12.5(c) is therefore finite whilst that for 12.3(c) was infinite. Consequently the string theory of gravity is expected to give finite amplitudes without any need for renormalization. A rigorous proof of this claim is as yet lacking. At laboratory energies the size of the string is so small that to all intents and purposes the particles behave as point-like, and the usual electroweak and strong theories are recovered. This limits the effective contact between string theory and experiment to the gravitational sector. None the less, for the first time there does exist a real possibility for the reconciliation of quantum theory and GR, the twin pillars of modern physics.

Appendix A

Variational methods

The general problem is to calculate what path between two points A and B in space–time makes the path integral

$$I = \int_A^B L\, d\tau$$

stationary. L is a function of the coordinates x^μ and their derivatives $q^\mu = dx^\mu/d\tau$ given in eqn (6.18). τ is the proper time or an equivalent scalar used to parametrize the path. The end-points are fixed points so that only variations on the route between them are to be considered. The requirement that the path is stationary can be written symbolically as

$$\delta I = \delta \int_A^B L\, d\tau = 0.$$

Suppose that the stationary path is known and that x^μ and q^μ are the coordinates and derivatives along this path. Next consider the small excursion from this path shown in Fig. A.1 over which these quantities become $x^\mu + \delta x^\mu$ and $q^\mu + \delta q^\mu$ respectively. Then the change in the path integral is

$$\delta I = \int_A^B \left(\frac{\partial L}{\partial x} \delta x + \frac{\partial L}{\partial q} \delta q \right) d\tau$$

where for clarity the superscript μ has been suppressed. Integrating the second term by parts gives

$$\left[\left(\frac{\partial L}{\partial q} \right) \delta x \right]_A^B - \int_A^B \delta x \left[\frac{d(\partial L/\partial q)}{d\tau} \right] d\tau.$$

Of these terms, the first vanishes because the end-points of the path are fixed so that $\delta x = 0$ at A and B. Thus

$$\delta I = \int_A^B \left[\frac{\partial L}{\partial x} - \frac{d(\partial L/\partial q)}{d\tau} \right] \delta x\, d\tau.$$

This variation must be zero for any choice of δx, which means that the term in square brackets must be identically zero along the whole path. Thus the stationary path has the differential equation (with the superscript μ restored)

$$\frac{\partial L}{\partial x^\mu} - \frac{d(\partial L/\partial q^\mu)}{d\tau} = 0 \tag{A.1}$$

If L is replaced by $L^{1/2}$, then the result is essentially unchanged.

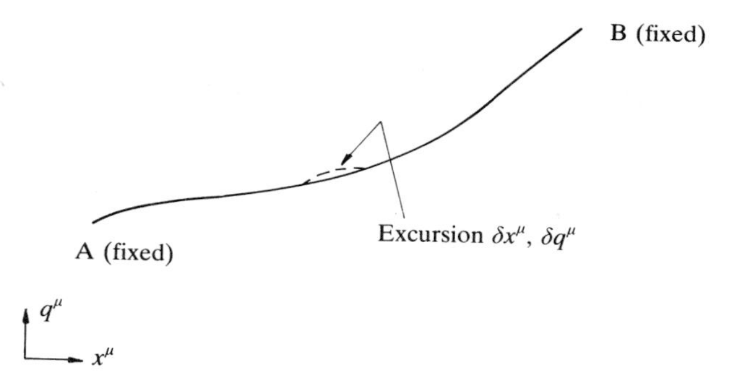

Fig. A.1

In order to obtain the geodesic equation from equation A.1 we make the substitution

$$L = g_{\alpha\beta} q^{\alpha} q^{\beta}.$$

Differentiating, and remembering that coordinates and velocities are independent quantities, we obtain

$$\frac{\partial L}{\partial x^{\mu}} = g_{\alpha\beta,\mu} q^{\alpha} q^{\beta}$$

$$\frac{\partial L}{\partial q^{\mu}} = g_{\alpha\mu} q^{\alpha} + g_{\mu\beta} q^{\beta}$$

$$\frac{d(\partial L/\partial q^{\mu})}{d\tau} = g_{\alpha\mu,\sigma} q^{\alpha} q^{\sigma} + g_{\mu\beta,\sigma} q^{\beta} q^{\sigma} + 2 g_{\mu\alpha} \frac{dq^{\alpha}}{d\tau}.$$

Substituting these values into eqn A.1 gives

$$g_{\alpha\beta,\mu} q^{\alpha} q^{\beta} - g_{\alpha\mu,\sigma} q^{\alpha} q^{\sigma} - g_{\mu\beta,\sigma} q^{\beta} q^{\sigma} - 2 g_{\mu\alpha} \frac{dq^{\alpha}}{d\tau} = 0.$$

The first three terms simplify using eqn (6.5)

$$-2\Gamma_{\mu\beta\alpha} q^{\alpha} q^{\beta} - 2 g_{\alpha\mu} \frac{dq^{\alpha}}{d\tau} = 0,$$

i.e.

$$g_{\mu\nu} \Gamma^{\nu}{}_{\beta\alpha} q^{\alpha} q^{\beta} + g_{\mu\nu} \frac{dq^{\nu}}{d\tau} = 0$$

where in the second term the equality of $g_{\mu\alpha}$ and $g_{\alpha\mu}$ has been used, and the repeated index has then been changed from α to ν. This reduces to

$$\Gamma^{\nu}{}_{\beta\alpha} q^{\alpha} q^{\beta} + \frac{dq^{\nu}}{d\tau} = 0.$$

Finally, expressing q^α in full, we have

$$\Gamma^{\nu}{}_{\beta\alpha}\frac{dx^\alpha}{d\tau}\frac{dx^\beta}{d\tau}+\frac{d^2x^\nu}{d\tau^2}=0, \tag{A.2}$$

which is the geodesic eqn (6.13).

If L is independent of one of the coordinates x^ν, then the corresponding component of eqn A.1 becomes much simpler

$$\frac{d(\partial L/\partial q^\nu)}{d\tau}=0, \tag{A.3}$$

while the remaining three component equations with $\mu \neq \nu$ retain the more general form of A.1. In this case $\partial L/\partial q^\nu$ is a constant of the motion, a result which is frequently used in Chapter 8. The reader may recognize that the variational methods employed here are closely related to the Lagrangian analysis of classical mechanics. In that case the quantity L is the Lagrangian $T - V$, where T is the kinetic energy and V the potential energy. In the cases of interest T is purely a quadratic function of the momentum so that, if V is independent of some coordinate x^μ, the result A.3 follows. This is the basis of the conservation laws of classical mechanics. If V, and hence L, is independent of the position coordinates, as is usually the case, then a conserved quantity is

$$\frac{\partial L}{\partial q^\mu}=\frac{\partial T}{\partial q^\mu}=\frac{\partial(q^2/2m)}{\partial q^\mu}=\frac{q_\mu}{m}.$$

In other words, the linear momentum is conserved. Similar calculations lead to the conservation laws of energy and angular momentum. The corresponding conserved quantities for Schwarzschild space–time emerge in Chapter 8 during the study of orbits around the Sun.

Appendix B

Curvature tensors

If a vector v^α is parallel transported around a closed path consisting of the displacements shown in Fig. 7.1 the total change in v^α is given by (eqn 7.2)

$$dv^\alpha = R^\alpha{}_{\beta\gamma\delta} v^\beta \, da^\delta \, db^\gamma,$$

where $R^\alpha{}_{\beta\gamma\delta}$ is the Riemann curvature tensor. The full expression for the Riemann curvature tensor is derived in Chapter 7 (eqn 7.3):

$$R^\alpha{}_{\beta\gamma\delta} = \Gamma^\alpha{}_{\beta\delta,\gamma} - \Gamma^\alpha{}_{\beta\gamma,\delta} + \Gamma^\alpha{}_{\sigma\gamma}\Gamma^\sigma{}_{\beta\delta} - \Gamma^\alpha{}_{\sigma\delta}\Gamma^\sigma{}_{\beta\gamma}.$$

On specializing to a frame in free fall the metric connections vanish and

$$R^\alpha{}_{\beta\gamma\delta} = \Gamma^\alpha{}_{\beta\delta,\gamma} - \Gamma^\alpha{}_{\beta\gamma,\delta} \tag{B.1}$$

Also in free fall,

$$R_{\alpha\beta\gamma\delta} = g_{\alpha\rho} R^\rho_{\beta\gamma\delta} = \Gamma_{\alpha\beta\delta,\gamma} - \Gamma_{\alpha\beta\gamma,\delta}.$$

This can be expressed in terms of the metric components using eqn (6.5):

$$R_{\alpha\beta\gamma\delta} = \frac{g_{\alpha\beta,\delta\gamma} + g_{\alpha\delta,\beta\gamma} - g_{\beta\delta,\alpha\gamma} - g_{\alpha\beta,\gamma\delta} - g_{\alpha\gamma,\beta\delta} + g_{\beta\gamma,\alpha\delta}}{2}$$

$$= \frac{g_{\alpha\delta,\beta\gamma} - g_{\beta\delta,\alpha\gamma} - g_{\alpha\gamma,\beta\delta} + g_{\beta\gamma,\alpha\delta}}{2}. \tag{B2}$$

Interchanging α with β, or γ with δ, reverses the sign of the last line so that

$$R_{\beta\alpha\gamma\delta} = R_{\alpha\beta\delta\gamma} = -R_{\alpha\beta\gamma\delta}. \tag{B.3}$$

Applying the symmetries

$$g_{\alpha\beta} = g_{\beta\alpha} \qquad \text{and} \qquad g_{\alpha\beta,\gamma\delta} = g_{\alpha\beta,\delta\gamma}$$

to (B.2) it follows that the Riemann curvature tensor is symmetric under the interchange $\alpha\beta \Leftrightarrow \gamma\delta$, i.e.

$$R_{\gamma\delta\alpha\beta} = R_{\alpha\beta\gamma\delta}. \tag{B.4}$$

Using the same symmetries it is easy to show that

$$R_{\alpha\beta\gamma\delta} + R_{\alpha\delta\beta\gamma} + R_{\alpha\gamma\delta\beta} = 0. \tag{B.5}$$

The above results have been derived in a freely falling frame, but because they are tensor relationships they must hold true in all frames. This is one more instance of the power of tensor methods. Let us write the Riemann tensor in a way that allows us to exploit the above symmetries, namely R_{xy} where x stands for $\alpha\beta$ and y stands for $\gamma\delta$. The symmetries of (B.3) and (B.4) mean that the only independent choice of subscripts is made by choosing the content of x and y. Neither the order of the contents ($\beta\alpha$ instead of $\alpha\beta$), nor the order of x and y (yx instead of xy) matters. There are six distinct choices for x and y, namely 01, 02, 03, 12, 13, and 23. The choices 00, 11, 22, and 33 give tensor components that vanish; for example using (B.3) we obtain

$$R_{00\gamma\delta} = -R_{00\gamma\delta},$$

and so this is zero. There are 21 independent ways of combining the six distinct choices. Equation (B.5) gives one further restriction when all the indices are different

$$R_{0123} + R_{0312} + R_{0231} = 0,$$

but collapses to the form of (B.3) or (B.4) in other cases. Thus there are only 20 independent components of the Riemann curvature tensor. There is one final useful identity between the components of the Riemann tensor. Differentiating this tensor, we obtain in a freely falling frame

$$R^{\alpha}{}_{\beta\gamma\delta,\mu} = \Gamma^{\alpha}{}_{\beta\delta,\gamma\mu} - \Gamma^{\alpha}{}_{\beta\gamma,\delta\mu}.$$

It follows that

$$R^{\alpha}{}_{\beta\gamma\delta,\mu} + R^{\alpha}{}_{\beta\delta\mu,\gamma} + R^{\alpha}{}_{\beta\mu\gamma,\delta} = 0.$$

This result can be extended to hold true in all frames if the simple space–time derivatives are replaced by covariant derivatives. The result is the Bianchi identities

$$R^{\alpha}{}_{\beta\gamma\delta;\mu} + R^{\alpha}{}_{\beta\delta\mu;\gamma} + R^{\alpha}{}_{\beta\mu\gamma;\delta} = 0. \tag{B.6}$$

The Ricci tensor is formed by contracting the Riemann tensor:

$$R_{\beta\delta} = R^{\alpha}{}_{\beta\alpha\delta} = g^{\alpha\sigma}R_{\sigma\beta\alpha\delta}.$$

In a freely falling frame this becomes

$$R_{\beta\delta} = g^{\alpha\sigma}\frac{g_{\sigma\delta,\beta\alpha} - g_{\beta\delta,\sigma\alpha} + g_{\beta\alpha,\sigma\delta} - g_{\sigma\alpha,\beta\delta}}{2} \tag{B.7}$$

The first term can be rewritten

$$g^{\alpha\sigma}g_{\sigma\delta,\beta\alpha} = g^{\sigma\alpha}g_{\alpha\delta,\beta\sigma} = g^{\alpha\sigma}g_{\alpha\delta,\beta\sigma} = g^{\alpha\sigma}g_{\delta\alpha,\sigma\beta}.$$

These equalities have made use of the symmetry of $g_{\alpha\delta}$ and the fact that the order of the differentiations is immaterial. Thus the sum of the first plus third terms in B.7 is symmetric under the interchange of β and δ. The second plus fourth terms are also symmetric under the same interchange; thus the Ricci tensor is symmetric under interchange of its two labels, i.e.

$$R_{\beta\delta} = R_{\delta\beta}.$$

This is a tensor equation and hence will hold true in all frames. The final contraction gives the Ricci scalar

$$R = g^{\beta\delta} R_{\beta\delta}.$$

Specializing to a frame in free fall gives

$$R = g^{\beta\delta} g^{\alpha\sigma} \frac{g_{\sigma\delta,\beta\alpha} - g_{\beta\delta,\sigma\alpha} + g_{\beta\alpha,\sigma\delta} - g_{\sigma\alpha,\beta\delta}}{2}.$$

Here the first and third terms are identical, as are the second and fourth, hence

$$R = g^{\beta\delta} g^{\alpha\sigma} (g_{\sigma\delta,\beta\alpha} - g_{\beta\delta,\sigma\alpha}).$$

The final tensor of interest is the Einstein tensor:

$$G_{\beta\delta} = R_{\beta\delta} - g_{\beta\delta} R/2.$$

This must be a symmetric tensor because both the metric tensor and the Ricci tensor are symmetric. It is also possible to show that the Einstein tensor is divergenceless. We commence this proof from the Bianchi identities of eqn (B.6):

$$R^{\alpha}{}_{\beta\gamma\delta;\mu} + R^{\alpha}{}_{\beta\delta\mu;\gamma} + R^{\alpha}{}_{\beta\mu\gamma;\delta} = 0.$$

Then contracting α with γ gives

$$R_{\beta\delta;\mu} + R^{\alpha}{}_{\beta\delta\mu;\alpha} - R_{\beta\mu;\delta} = 0.$$

Raising the index β and contracting this with δ gives

$$R^{\beta}{}_{\beta;\mu} - R^{\alpha}{}_{\mu;\alpha} - R^{\beta}{}_{\mu;\beta} = 0,$$

which reduces to

$$(R - 2R^{\alpha}{}_{\mu})_{;\alpha} = 0,$$

i.e.

$$G^{\alpha}{}_{\mu;\alpha} = 0.$$

This is the desired result, showing the Einstein tensor to be a divergenceless symmetric second-rank tensor constructed solely from contractions of the Riemann curvature tensor.

In this final section we shall calculate the metric connections and the curvature tensors for the case of a two-dimensional spherical surface. This surface is simple in having the same curvature everywhere. The starting point is the definition of interval in polar coordinates:

$$ds^2 = r^2\, d\theta^2 + r^2 \sin^2\theta\, d\varphi^2.$$

Then with $x^1 = \theta$ and $x^2 = \varphi$,

$$g_{11} = r^2, \qquad g_{22} = r^2 \sin^2\theta$$

$$g^{11} = r^{-2} \qquad g^{22} = \frac{1}{r^2 \sin^2\theta}.$$

In turn the only non-zero derivatives are

$$g_{22,1} = 2r^2 \sin\theta \cos\theta$$

and

$$g_{22,11} = 2r^2(\cos^2\theta - \sin^2\theta)$$

The non-zero elements of the metric connections are

$$\Gamma^2{}_{21} = \frac{g^{22}g_{22,1}}{2} = \frac{\cos\theta}{\sin\theta}$$

$$\Gamma^2{}_{12} = \frac{g^{22}g_{22,1}}{2} = \frac{\cos\theta}{\sin\theta}$$

$$\Gamma^1{}_{22} = \frac{g^{11}(-g_{22,1})}{2} = -\sin\theta\cos\theta.$$

Their derivatives are

$$\Gamma^2{}_{21,1} = \Gamma^2{}_{12,1} = -1/\sin^2\theta$$

$$\Gamma^1{}_{22,1} = \sin^2\theta - \cos^2\theta.$$

One non-zero component of the Riemann tensor is

$$R^1{}_{212} = \Gamma^1{}_{22,1} - \Gamma^1{}_{21,2} + \Gamma^\sigma{}_{22}\Gamma^1{}_{\sigma 1} - \Gamma^\sigma{}_{21}\Gamma^1{}_{\sigma 2}$$

$$= \sin^2\theta - \cos^2\theta - \frac{\cos\theta}{\sin\theta}(-\sin\theta\cos\theta)$$

$$= \sin^2\theta.$$

It is clear that only R_{1212} and rearrangements such as R_{2121} are not zero. Explicitly

$$R_{1212} = r^2 \sin^2\theta \qquad \text{and} \qquad R^2{}_{121} = g^{22}R_{2121} = 1.$$

The Gaussian curvature given by Gauss's 'Excellent' Theorem is

$$K = \frac{R_{1212}}{g} = \frac{R_{1212}}{g_{11}g_{22}}$$

with the opposite sign from (7.8) because here the components of the metric tensor are all positive. It then follows that $K = 1/r^2$. The Ricci tensor has two non-zero components:

$$R_{11} = R^2{}_{121} = 1 \qquad \text{and} \qquad R_{22} = R^1{}_{212} = \sin^2\theta.$$

The Ricci scalar is given by

$$R = g^{11}R_{11} + g^{22}R_{22} = 2/r^2,$$

while corresponding components of the Einstein tensor are

$$G_{11} = R_{11} - g_{11}R/2 = 1 - r^2/r^2 = 0$$

$$G_{22} = R_{22} - g_{22}R/2 = 0.$$

Thus the Einstein curvature tensor vanishes for the sphere.

Appendix C

The equation for the metric connection

First a symmetry property of the metric connections must be derived. Consider a scalar potential φ which varies smoothly in space–time, and from it construct the force

$$F_{\mu} = \varphi_{,\mu} = \frac{\partial \varphi}{\partial x^{\mu}}.$$

The covariant derivative of a covector F_{μ} is

$$\frac{DF_{\mu}}{Dx^{\rho}} = \varphi_{,\mu\rho} - F_{\sigma}\Gamma^{\sigma}{}_{\mu\rho}. \tag{C.1}$$

Note that the sign of the second term is negative for covector and positive for vector components. Then in a freely falling frame

$$\frac{DF_{\mu}}{Dx^{\rho}} = \varphi_{,\mu\rho}.$$

$\varphi_{,\mu\rho}$ is symmetric under the interchange $\mu \leftrightarrow \rho$; hence DF_{μ}/Dx^{ρ} must also be symmetric under the interchange. This property remains true in all frames for $\varphi_{,\mu\rho}$ and DF_{μ}/Dx^{ρ}. Accordingly, using eqn (C.1), we see that the connections $\Gamma^{\sigma}_{\rho\mu}$ must share this symmetry:

$$\Gamma^{\sigma}_{\rho\mu} = \Gamma^{\sigma}_{\mu\rho} \qquad \text{and} \qquad \Gamma_{\sigma\rho\mu} = \Gamma_{\sigma\mu\rho}. \tag{C.2}$$

Now rewriting (6.4) and repeating it with indices permuted gives

$$g_{\mu\nu,\rho} = \Gamma_{\nu\mu\rho} + \Gamma_{\mu\nu\rho} \tag{C.3}$$

$$g_{\rho\mu,\nu} = \Gamma_{\mu\rho\nu} + \Gamma_{\rho\mu\nu}$$

$$g_{\nu\rho,\mu} = \Gamma_{\rho\nu\mu} + \Gamma_{\nu\rho\mu}.$$

These last two equations can be converted using eqn (C.2) to give

$$g_{\rho\mu,\nu} = \Gamma_{\mu\nu\rho} + \Gamma_{\rho\mu\nu} \tag{C.4}$$

and

$$g_{\nu\rho,\mu} = \Gamma_{\rho\mu\nu} + \Gamma_{\nu\mu\rho}. \tag{C.5}$$

Taking the combination C.3 − C.4 + C.5 yields

$$2\Gamma_{\nu\mu\rho} = g_{\mu\nu,\rho} - g_{\rho\mu,\nu} + g_{\nu\rho,\mu}. \tag{C.6}$$

The metric connections are unique functions, as well as being symmetric under the interchange of the last two indices ($\mu \leftrightarrow \rho$).

Appendix D

The Schwarzschild metric

The spherically symmetric metric satisfying Einstein's equation for empty space is the Schwarzschild metric. This is straightforward to demonstrate using the additional assumption that the metric is static. The most general spherically symmetric metric is

$$\mathrm{d}s^2 = ac^2\,\mathrm{d}t^2 - b\,\mathrm{d}r^2 - 2e\,\mathrm{d}t\,\mathrm{d}r - r^2\,\mathrm{d}\Omega^2$$

where a, b, and e are functions of r and t only. A new time coordinate t' can be chosen for which

$$c\,\mathrm{d}t' = \left(a^{1/2}\,\mathrm{d}t - \frac{e}{a^{1/2}}\,\mathrm{d}r\right) f(t, r)$$

where f is the integrating factor needed to make the right-hand side a perfect differential. Then

$$\mathrm{d}s^2 = Ac^2\,\mathrm{d}t'^2 - B\,\mathrm{d}r^2 - r^2\,\mathrm{d}\Omega^2.$$

If the metric is *static* then A and B depend on r only. Dropping the prime gives

$$\mathrm{d}s^2 = Ac^2\,\mathrm{d}t^2 - B\,\mathrm{d}r^2 - r^2\,\mathrm{d}\Omega^2 \tag{D.1}$$

with metric components

$$g_{00} = A, \qquad g_{11} = -B, \qquad g_{22} = -r^2, \qquad g_{33} = -r^2 \sin^2\theta.$$

The metric is diagonal so that $g^{00} = 1/g_{00}$ etc. By using the definition of eqn (6.5) the non-zero components of the metric connection can be obtained:

$$\Gamma^0_{01} = \Gamma^0_{10} = A'/2A$$

$$\Gamma^1_{00} = A'/2B, \qquad \Gamma^1_{11} = B'/2B$$

$$\Gamma^1_{22} = -r/B, \qquad \Gamma^1_{33} = -r\sin^2\theta/B$$

$$\Gamma^2_{12} = \Gamma^2_{21} = \Gamma^3_{13} = \Gamma^3_{31} = 1/r$$

$$\Gamma^2_{33} = -\sin\theta\cos\theta \qquad \Gamma^3_{32} = \Gamma^3_{23} = \cot\theta.$$

Here the prime denotes differentiation with respect to r. Next, using the definition of the Ricci tensor,

$$R_{\beta\delta} = R^{\alpha}_{\beta\alpha\delta}$$

and eqn (7.3) for the Riemann curvature tensor, we obtain

$$R_{00} = \frac{A''}{2B} - \frac{A'}{4B}\left(\frac{A'}{A} + \frac{B'}{B}\right) + \frac{A'}{rB} \tag{D.2a}$$

$$R_{11} = -\frac{A''}{2A} + \frac{A'}{4A}\left(\frac{A'}{A} + \frac{B'}{B}\right) + \frac{B'}{rB} \tag{D.2b}$$

$$R_{22} = 1 - \frac{r}{2B}\left(\frac{A'}{A} - \frac{B'}{B}\right) - \frac{1}{B} \tag{D.2c}$$

$$R_{33} = R_{22}\sin^2\theta. \tag{D.2d}$$

Using eqns (D.2(a)) and (D.2(b)) gives

$$\frac{R_{00}}{A} + \frac{R_{11}}{B} = \frac{(A'/A + B'/B)}{rB}. \tag{D.3}$$

Now in empty space Einstein's equation reduces to

$$R_{\mu\nu} = 0. \tag{D.4}$$

Therefore eqn (D.3) becomes

$$A'/A + B'/B = 0$$

whence

$$AB = \text{constant}.$$

At points remote from the source both A and B tend to unity, so that the constant is unity. Therefore

$$B = 1/A.$$

Substituting this value for B in eqns (D.2(c)) and making use of (D.4) again gives

$$1 - rA' - A = 0.$$

Integration with respect to r gives

$$rA - r = \text{constant}.$$

However, experiments on the gravitational red shift fix the asymptotic form of A to be $1 - 2GM/rc^2$. Therefore the constant is $-2GM/c^2$, and so the general form for A is

$$A = 1 - \frac{2GM}{rc^2},$$

which is identical with the asymptotic form. Then, since $B = 1/A$, the complete metric equation is

$$\mathrm{d}s^2 = \left(1 - \frac{2GM}{rc^2}\right)c^2\,\mathrm{d}t^2 - \frac{\mathrm{d}r^2}{1 - 2GM/rc^2} - r^2\,\mathrm{d}\Omega^2. \tag{D.5}$$

Next we evaluate tensors for the Schwarzschild metric. For convenience we put

$$m = 2GM/c^2, \qquad Z = 1 - m/r.$$

Then

$$g_{00} = Z, \qquad g_{11} = -1/Z, \qquad g_{22} = -r^2, \qquad g_{33} = -r^2 \sin^2\theta.$$

The non-zero metric connections are

$$\Gamma^1_{00} = mZ/2r^2, \qquad \Gamma^0_{01} = \Gamma^0_{10} = m/2r^2Z$$

$$\Gamma^2_{12} = \Gamma^2_{21} = 1/r, \qquad \Gamma^1_{22} = -rZ$$

$$\Gamma^1_{11} = -m/2r^2Z, \qquad \Gamma^1_{33} = -rZ\sin^2\theta$$

$$\Gamma^3_{31} = \Gamma^3_{13} = 1/r, \qquad \Gamma^2_{33} = -\sin\theta\cos\theta$$

$$\Gamma^3_{32} = \Gamma^3_{23} = \frac{\cos\theta}{\sin\theta}.$$

The non-zero elements of the Riemann curvature tensor are

$$R^3_{131} = -m/2r^3Z, \qquad R^1_{010} = -mZ/r^3$$

$$R^1_{212} = -m/2r, \qquad R^2_{323} = m\sin^2\theta/r$$

$$R^0_{303} = -m\sin^2\theta/2r, \qquad R^0_{202} = -m/2r$$

plus components which are related by symmetry operations, e.g.

$$-R_{3113} = -R_{1331} = R_{1313} = R_{3131}$$

where

$$R_{3131} = g_{33}R^3_{131} \qquad \text{etc.}$$

The components of the Ricci tensor

$$R_{\mu\nu} = R^\alpha_{\mu\alpha\nu}$$

all vanish, as they must if the metric satisfies Einstein's equation in empty space–time. Finally we obtain the curvature of two-dimensional surfaces in space–time. Take the surface defined by geodesics, which at the origin lie in directions such that the polar angle θ is $\pi/2$ and also have t constant. The Gaussian curvature of this surface is

$$K_{r\varphi} = -\frac{R_{3131}}{g_{11}g_{33}} = -\frac{R^3_{131}}{g_{11}} = -\frac{m}{2r^3},$$

i.e.

$$K_{r\varphi} = -\frac{GM}{r^3c^2}.$$

This has finally checked the consistency of the choice of curvature expressed in eqn (4.8).

Appendix E

The energy flow in gravitational waves

In the linearized approximation appropriate to regions of space–time where gravitational curvature is small, the metric is written

$$g_{\alpha\beta} = \eta_{\alpha\beta} + h_{\alpha\beta}; \qquad h_{\alpha\beta} \ll 1 \tag{10.2}$$

The stress-energy tensor for gravitational waves was shown in Section 10.1 (eqn 10.16) to be

$$t_{\alpha\beta} = -\frac{c^4}{8\pi G} G^{(2)}_{\alpha\beta} \tag{E.1}$$

where only terms quadratic in $h_{\alpha\beta}$ are retained on the right-hand side. This stress-energy tensor will now be calculated for a plane transverse–traceless wave with (+) polarization using Cartesian coordinates. Thus

$$h_+ = h_{11} = -h_{22} = h\cos[k(x^0 - x^3)]. \tag{E.2}$$

The only components of the metric that do not vanish are

$$\begin{aligned} g_{11} &= -1 + h_+, & g^{11} &= -1 - h_+ \\ g_{22} &= -1 - h_+, & g^{22} &= -1 + h_+ \\ g_{00} &= g^{00} = 1, & g_{33} &= g^{33} = -1. \end{aligned}$$

Using eqn (6.5) we obtain

$$2\Gamma^1_{01} = 2g^{11}\Gamma_{101} = g^{11}g_{11,0} = -h_{+,0} - h_+h_{+,0},$$

where the product

$$h_+h_{+,0} \propto \sin[2k(x^0 - x^3)].$$

One later step in calculating energy flow requires us to take the average of the stress-energy tensor over a region of space–time larger than several complete waves. With this averaging procedure any contribution from the above product vanishes:

$$\langle h_+h_{+,0}\rangle = 0. \tag{E.3}$$

Therefore we can ignore such terms from here on. Then

$$\Gamma^1_{10} = \Gamma^1_{01} = \Gamma^0_{11} = -h_{+,0}/2$$

and

$$\Gamma^1_{13} = \Gamma^1_{31} = -\Gamma^3_{11} = -h_{+,3}/2 = +h_{+,0}/2.$$

Similar formulae hold with the suffix 1 replaced everywhere by 2. Referring to the expression for the Riemann tensor (eqn 7.3) shows that if only the terms quadratic in h_+ are retained we have

$$R^{(2)\alpha}{}_{\beta\gamma\delta} = \Gamma^\alpha_{\sigma\gamma}\Gamma^\sigma_{\beta\delta} - \Gamma^\alpha_{\sigma\delta}\Gamma^\sigma_{\beta\gamma}.$$

Relevant non-zero components of this Riemann tensor are

$$\begin{aligned} -R^{(2)1}{}_{010} &= -R^{(2)2}{}_{020} = h^2_{+,0}/4 \\ -R^{(2)0}{}_{101} &= R^{(2)3}{}_{131} = h^2_{+,0}/4 \\ -R^{(2)1}{}_{313} &= -R^{(2)2}{}_{323} = h^2_{+,0}/4, \end{aligned} \tag{E.4}$$

whence it follows that the second-order components of the Ricci tensor are

$$-R^{(2)}_{00} = R^{(2)}_{30} = R^{(2)}_{03} = -R^{(2)}_{33} = h^2_{+,0}/2 \tag{E.5}$$

while for the Ricci scalar

$$R^{(2)} = 0.$$

Therefore the Einstein tensor components are identical with the Ricci tensor components:

$$-G^{(2)}_{00} = G^{(2)}_{30} = G^{(2)}_{03} = -G^{(2)}_{33} = h^2_{+,0}/2. \tag{E.6}$$

Finally, substituting these results in eqn (E.1) we have the components of the stress-energy tensor of the gravitational wave, for example

$$t_{00} = \frac{c^4}{16\pi G} h^2_{+,0}.$$

Now the energy in a wave cannot be localized because only relative displacements have physical significance. It is also not clear either whether the energy is located in a valley or a peak. Hence it is necessary to average over several complete cycles giving

$$t_{00} = \frac{c^4}{16\pi G} \langle h^2_{+,0} \rangle$$

where the angular brackets refer to expectation values. Converting to a time derivative, we obtain

$$t_{00} = \left(\frac{c^2}{16\pi G}\right) \langle \dot{h}^2_+ \rangle.$$

If both polarization states are present this result generalizes to

$$t_{00} = \frac{c^2}{16\pi G} \langle \dot{h}^2_+ + \dot{h}^2_\times \rangle. \tag{E.7}$$

It follows that the energy flux is

$$F = ct_{00} = \frac{c^3}{16\pi G}\langle \dot{h}_+^2 + \dot{h}_\times^2 \rangle \tag{E.8}$$

meaning the energy crossing unit area per unit time (kg s^{-3}). Equation (E.8) can be rewritten as

$$F = \frac{c^3}{32\pi G}\langle \dot{h}_{11}^2 + \dot{h}_{12}^2 + \dot{h}_{21}^2 + \dot{h}_{22}^2 \rangle,$$

i.e.

$$F = \frac{c^3}{32\pi G}\langle \dot{h}_{ij}^{TT}\dot{h}_{ij}^{TT} \rangle, \tag{E.9}$$

where we make explicit that we are working in the transverse–traceless gauge by adding the superscript TT.

Appendix F

Radiation from a nearly Newtonian source

A Newtonian source is one in which the curvature and strain of space–time are small, and in which the motion of matter is non-relativistic. Then the metric tensor can be written approximately as

$$g_{\alpha\beta} = \eta_{\alpha\beta} + h_{\alpha\beta}, \qquad h_{\alpha\beta} \ll 1.$$

When a transverse–traceless gauge is chosen, Einstein's equation reduces to

$$\frac{\partial^2 h_{\alpha\beta}}{\partial x^\mu \partial x_\mu} = \frac{16\pi G T_{\alpha\beta}}{c^4} \tag{F.1}$$

where $T_{\alpha\beta}$ is the stress-energy tensor which vanishes outside the source. The details of the calculation follow the steps leading to eqn (10.4). Far from the source the solution of (F.1) is

$$h_{\alpha\beta}(t) = \frac{4G}{c^4} \int T_{\alpha\beta}\left(t - \frac{r}{c}\right)\frac{dV}{r}$$

where the integral is performed over the source and r is the distance from the source to the point where $h_{\alpha\beta}$ is required. This solution closely resembles the retarded solution for the electromagnetic field of a remote source. There is a time delay r/c because the wave has to travel out a distance r. Let y_μ be the coordinates of a representative point in the source. Then

$$h_{\alpha\beta}(t) = \frac{4G}{rc^4} \int T_{\alpha\beta}\left(t - \frac{r}{c}\right) \mathrm{d}^3 y$$

provided that $y \ll r$. For the assumed non-relativistic source

$$T_{00} \approx \rho_0 c^2 + \rho_0 v^2/2$$

$$T_{ij} \approx \rho_0 v_i v_j$$

where ρ_0 is the rest density and v is the velocity ($\ll c$). Thus

$$\frac{\partial^2}{\partial t^2}\left(\int T_{00} y_i y_j \,\mathrm{d}^3 y\right) \approx 2c^2 \int T_{ij}\,\mathrm{d}^3 y$$

where we have made use of the fact that all accelerations are small. Therefore

$$h_{ij} = \frac{2G}{rc^4}\frac{\partial^2}{\partial t^2}\left(\int \rho_0 y_i y_j \,\mathrm{d}^3 y\right).$$

We drop the time arguments in order to simplify the presentation. The integrand in brackets is a quadrupole mass moment of the source:

$$I_{ij} = \int \rho_0 y_i y_j \, \mathrm{d}^3 y.$$

The choice we make for h_{ij} is that it is transverse and traceless, and so it follows that the quadrupole moment must be similarly rendered traceless and transverse. A conventional starting point is to take the reduced quadrupole moment

$$\not{I}_{ij} = \int T_{00}(y_i y_j - \delta_{ij} y^2{}_k) \, \mathrm{d}^3 y. \tag{F.2}$$

The Kronecker delta δ_{ij} has the value $+1$ when $i = j$ and is zero otherwise. Now the component of a vector $\boldsymbol{d}$ transverse to a unit vector $\boldsymbol{n}$ is given by

$$d_j^{\mathrm{T}} = P_{jl} d_l$$

where $P_{jl} = (\delta_{jl} - n_j n_l)$. In the present case $\boldsymbol{n}$ is a unit vector $\boldsymbol{r}/r$ pointing from the source. Similarly a transverse version of $\not{I}_{ij}$ is

$$I_{ij}^{\mathrm{T}} = P_{il} P_{jm} \not{I}_{lm}.$$

However, the trace of this moment does not vanish:

$$\begin{aligned} I_{jj}^{\mathrm{T}} &= P_{jl} P_{jm} \not{I}_{lm} = (\delta_{lm} - n_l n_m) \not{I}_{lm} \\ &= P_{lm} \not{I}_{lm} \\ &= P_{ij} P_{lm} \not{I}_{lm} / 2 \end{aligned}$$

where we have used the identity

$$P_{jj} = \delta_{jj} - n_j n_j = 2.$$

The appropriate transverse–traceless quadrupole moment of the source is therefore

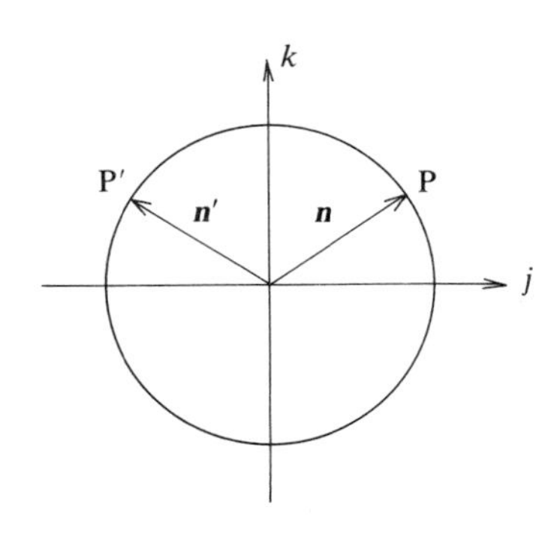

Fig. F.1

$$I^{\mathrm{TT}}_{ij} = P_{il}P_{jm}\not{I}_{lm} - P_{ij}P_{lm}\not{I}_{lm}/2. \tag{F.3}$$

Thus

$$h_{ij} = \frac{2G}{rc^4}\ddot{I}^{\mathrm{TT}}_{ij}. \tag{F.4}$$

Using earlier results from Appendix E we can calculate the energy flux at points distant from the source and also the luminosity of the source. Equation (E.9) gives the flux

$$\begin{aligned} F &= \frac{c^3}{32\pi G}\langle \dot{h}_{ij}\dot{h}_{ij}\rangle \\ &= \frac{G}{8\pi r^2 c^5}\langle \dddot{I}^{\mathrm{TT}}_{ij}\dddot{I}^{\mathrm{TT}}_{ij}\rangle \end{aligned} \tag{F.5}$$

Integrating over a sphere at a distance r from the source gives the total flux, and hence the luminosity of the source is

$$L = r^2 \int F \, \mathrm{d}\Omega.$$

Writing the product of transverse–traceless quadrupole moments in terms of the reduced moments gives

$$F = \frac{G}{16\pi r^2 c^5}\langle(2\dddot{\not{I}}_{ij}\dddot{\not{I}}_{ij} - 4n_j n_k \dddot{\not{I}}_{ij}\dddot{\not{I}}_{ki} + n_i n_j n_k n_l \dddot{\not{I}}_{ij}\dddot{\not{I}}_{kl})\rangle.$$

It is easy to see that the integral

$$L_2(j, k) = \int n_j n_k \, \mathrm{d}\Omega$$

vanishes if $j \neq k$. Figure F.1 shows a slice through the sphere of integration at constant x_l, where l is the label of the remaining space coordinate ($l \neq k$, $l \neq i$). The value of n_k at P' is opposite to its value at P but n_j is the same; consequently the contributions near P and P' cancel. Similarly all contributions to the integral cancel. When $j = k$, this does not happen. For example, if $j = k = 3$ we have

$$L_2(3, 3) = \int_{-\pi}^{+\pi}\int_0^{2\pi} \cos^2\theta \sin\theta \, \mathrm{d}\theta \, \mathrm{d}\varphi = 4\pi/3.$$

In summary $L_2(j, k) = 4\pi\delta_{jk}/3$. A similar analysis shows that

$$\int n_i n_j n_k n_l \, \mathrm{d}\Omega = \frac{4\pi}{15}(\delta_{ij}\delta_{kl} + \delta_{ik}\delta_{jl} + \delta_{il}\delta_{jk}).$$

Finally, using these results in the integral over energy flux gives

$$L = (G/5c^5)\langle \dddot{\not{I}}_{ij}\dddot{\not{I}}_{ij}\rangle. \tag{F.6}$$

Appendix G

The deflection of light in the Schwarzschild metric

In this appendix the deflection of light is recalculated in such a way that the contributions of frame and time distortion are separated. The calculation proceeds using isotropic coordinates: the new radial coordinate r is given by

$$r = \bar{r}\left(1 + \frac{GM}{2\bar{r}c^2}\right)^2. \tag{G.1}$$

Thus

$$\begin{aligned} \mathrm{d}r &= \mathrm{d}\bar{r}\left(1 - \frac{G^2M^2}{4\bar{r}^2c^4}\right) \\ &= \mathrm{d}\bar{r}\left(1 - \frac{GM}{2\bar{r}c^2}\right)\left(1 + \frac{GM}{2\bar{r}c^2}\right) \end{aligned} \tag{G.2}$$

and

$$\left(1 - \frac{2GM}{rc^2}\right) = \frac{(1 - GM/2\bar{r}c^2)^2}{(1 + GM/2\bar{r}c^2)^2}. \tag{G.3}$$

The metric equation for Schwarzschild space specialized to the case of photons travelling in the equatorial plane ($\theta = \pi/2$, $\mathrm{d}s^2 = 0$) becomes

$$0 = c^2\,\mathrm{d}t^2\,\frac{(1 - GM/2\bar{r}c^2)^2}{(1 + GM/2\bar{r}c^2)^2} - \left(1 + \frac{GM}{2\bar{r}c^2}\right)^4(\mathrm{d}\bar{r}^2 + \bar{r}^2\,\mathrm{d}\varphi^2).$$

Light therefore has the same velocity in all directions, namely $c(1 - GM/2\bar{r}c^2)/(1 + GM/2\bar{r}c^2)^3$. Hence the coordinates are called isotropic. Then

$$0 = c^2\,\mathrm{d}t^2\,\frac{(1 - GM/2\bar{r}c^2)^2}{(1 + GM/2\bar{r}c^2)^2} - \left(1 + \frac{GM}{2\bar{r}c^2}\right)^4(\mathrm{d}\bar{x}^2 + \mathrm{d}\bar{y}^2)$$

where $\bar{x}$ and $\bar{y}$ are local Cartesian coordinates in the equatorial plane. To first order in GM/rc^2 this is

$$0 = c^2\,\mathrm{d}t^2\left(1 - \frac{2GM}{\bar{r}c^2}\right) - \left(1 + \frac{2GM}{\bar{r}c^2}\right)(\mathrm{d}\bar{x}^2 + \mathrm{d}\bar{y}^2). \tag{G.4}$$

For simplicity the light beam is assumed to be emitted parallel to the $\bar{x}$ axis. Then the $\bar{y}$ component of acceleration is given by the null geodesic equation

$$\frac{d^2\bar{y}}{d\lambda^2} + \Gamma^2{}_{\mu\nu}\frac{dx^\mu}{d\lambda}\frac{dx^\nu}{d\lambda} = 0 \tag{G.5}$$

where λ is some path-length parameter. The deflection is slight, so that the only non-negligible differentials are $d\bar{x}/d\lambda$ and $dt/d\lambda$. Also $\Gamma^2{}_{10}$ and $\Gamma^2{}_{01}$ vanish because the metric is diagonal. Thus eqn (G.5) reduces to

$$\frac{d^2\bar{y}}{d\lambda^2} + \Gamma^2{}_{00}c^2\left(\frac{dt}{d\lambda}\right)^2 + \Gamma^2{}_{11}\left(\frac{d\bar{x}}{d\lambda}\right)^2 = 0. \tag{G.6}$$

The trajectory of the light remains close to the x direction; hence from eqn (G.4)

$$\frac{dt}{d\lambda} = \frac{dt}{d\bar{x}}\frac{d\bar{x}}{d\lambda} = \frac{d\bar{x}}{d\lambda}\frac{1 + 2GM/\bar{r}c^2}{c}$$

to first order in $GM/\bar{r}c^2$. Substituting this value into eqn (G.6) and dividing through by $(d\bar{x}/d\lambda)^2$ gives

$$\frac{d^2\bar{y}}{d\bar{x}^2} + \Gamma^2{}_{00}\left(1 + \frac{2GM}{\bar{r}c^2}\right)^2 + \Gamma^2{}_{11} = 0 \tag{G.7}$$

Using again the fact that the metric is diagonal, we obtain

$$\Gamma^2{}_{00} = g^{22}\Gamma_{200} = \frac{-\Gamma_{200}}{1 + 2GM/\bar{r}c^2}$$

and

$$\Gamma^2{}_{11} = g^{22}\Gamma_{211} = \frac{-\Gamma_{211}}{1 + 2GM/\bar{r}c^2}$$

From eqn (6.5),

$$\Gamma^2{}_{00} = \frac{g_{00,2}}{2(1 + 2GM/\bar{r}c^2)}.$$

Now

$$g_{00,2} = \frac{\partial g_{00}}{\partial \bar{y}} = \frac{2GM}{\bar{r}^2c^2}\frac{\partial \bar{r}}{\partial \bar{y}} = \frac{2GM\bar{y}}{\bar{r}^3c^2}.$$

Thus

$$\Gamma^2{}_{00} = \frac{GM\bar{y}}{\bar{r}^3c^2(1 + 2GM/\bar{r}c^2)}$$

and then to order $GM/\bar{r}c^2$

$$\Gamma^2{}_{00} = \frac{GM\bar{y}}{\bar{r}^3c^2}. \tag{G.8}$$

Similarly, to the same approximation,

$$\Gamma^2{}_{11} = \frac{GM\bar{y}}{\bar{r}^3 c^2}. \tag{G.9}$$

Inserting these values for the metric connections in the geodesic equation (G.7) and neglecting terms of higher order than $GM/\bar{r}c^2$ gives

$$\frac{d^2\bar{y}}{d\bar{x}^2} = \frac{2GM\bar{y}}{\bar{r}^3 c^2}.$$

The angle change on passing the Sun is

$$\Delta\Theta = \int_{-\infty}^{+\infty} \frac{d^2\bar{y}}{d\bar{x}^2}\, d\bar{x}.$$

If b is the impact parameter, $\bar{y} \approx b$ and

$$\Delta\Theta = \frac{4GM}{c^2}\int_0^{\infty} \frac{b\, d\bar{x}}{(\bar{x}^2 + b^2)^{3/2}}.$$

Now put $\tan\varphi$ equal to $\bar{x}/b$ and then $d\bar{x} = b\sec^2\varphi\, d\varphi$ and

$$\Delta\Theta = \frac{4GM}{c^2 b}\int_0^{\pi/2} \cos\varphi\, d\varphi = \frac{4GM}{bc^2}$$

which reproduces eqn (8.14) in Section 8.2. The crucial feature revealed here is that the contributions due to time distortion (eqn (G.8)) and frame distortion (eqn (G.9)) are equal. Where the full GR treatment yields a deflection of 1.75 arcsec for light passing the limb of the Sun, a prediction based on the equivalence principle alone would neglect the frame distortion and give exactly half this deflection.

Appendix H

Table of physical constants and parameters

Gravitational constant, $G = 6.6726 \times 10^{-11}$ kg^{-1} m^3 s^{-2}
Speed of light, $c = 2.99792 \times 10^8$ m s^{-1}
Mass of Earth, $M_\oplus = 5.98 \times 10^{24}$ kg
Mean radius of Earth, $R_\oplus = 6.37 \times 10^6$ m
Mean radius of Earth's orbit, 1.50×10^{11} m
Mass of Sun, $M_\odot = 1.99 \times 10^{30}$ kg
Radius of Sun, $R_\odot = 6.96 \times 10^8$ m
Hubble's constant, $H_0 = 50$–100 km s^{-1} Mpc^{-1}
Deceleration parameter, $-1.3 < q_0 < +2.0$
Cosmological constant, $|\Lambda| < 10^{-52}$ m^{-2}
Density of Universe, $10^{-27} \lesssim \rho_0 \lesssim 10^{-26}$ kg m^{-3}
Diameter of visible Galaxy, ≈ 30 kpc
Distance to Virgo cluster, ≈ 15 Mpc
Luminosity factor, $L_0 = c^5/G = 3.63 \times 10^{52}$ W
1 year $= 3.156 \times 10^7$ s
1 light-year $= 9.46 \times 10^{15}$ m
1 parsec $= 3.09 \times 10^{16}$ m $= 3.26$ light-years
Planck's constant, $h = 6.626 \times 10^{-34}$ J s
Electron charge, $e = 1.602 \times 10^{-19}$ C
Electron mass, $m = 9.109 \times 10^{-31}$ kg
Proton mass, $m_p = 1.673 \times 10^{-27}$ kg
Avogadro's number, $N_A = 6.022 \times 10^{23}$ mol^{-1}
Boltzmann constant, $k = 1.381 \times 10^{-23}$ J K^{-1}
1 Galileo (Gal) $= 10^{-2}$ m s^{-2}

Appendix I

Questions

Q1.1 Calculate the four-momentum of a proton (mass 938 MeV/c^2) moving with velocity $0.8c$ (a) along the x axis and (b) in a direction making equal angles with all the x, y, and z axes.

Q1.2 Calculate whether the following space–time intervals are space-like, time-like, or light-like: (1, 3, 0, 0); (3, 3, 0, 0); (3, −3, 0, 0); (−3, 3, 0, 0); (0, 3, 0, 0); (3, 1, 0, 0).

Q1.3 What is the proper time interval between the emission of neutrinos from supernova 1987A in the large Magellanic cloud 170 light-years from Earth and their arrival on Earth in February 1987.

Q2.1 Show that in the Rebka and Pound experiment the expected gravitational red shift is 2.46×10^{-15}.

Q2.2 Calculate the gravitational red shift in wavelength for the 769.9 nm potassium line emitted from the Sun's surface.

Q3.1 Calculate the approximate rotation in the horizontal plane of a vector if it is parallel transported around the periphery of the contiguous 48 states of the USA.

Q3.2 A two-dimensional toroidal surface has these dimensions: the mean diameter is 20 m and the radius of the circular cross-section is 2 m. Calculate the Gaussian curvature of the surface at the inner and outer edges of the torus. Where is the Gaussian curvature of this surface infinite?

Q3.3 What is the metric equation of an isotropic homogeneous three-dimensional space with curvature $-10^{-2}\ \text{m}^{-2}$? What is the area of a spherical surface whose measured distance from the central point is 10 m?

Q4.1 A ball is thrown vertically upward and returns to the thrower's hand in time t. Take local coordinates with the x axis upward and the time starting at zero as the ball is thrown. What are the four-vector coordinates of the ball at its highest point, and when it returns to the thrower? What is the curvature of this path in space–time?

Q4.2 A star of the same mass as the Sun has radius 3 km. Calculate the metric equation for space–time near the surface.

Q4.3 Write down the metric equation for an equatorial line in Schwarzschild space at constant time ($\theta = \pi/2, \varphi = 0, t$ constant). You should obtain

$$ds^2 = \frac{dr^2}{1 - r_0/r} \tag{1}$$

where $r_0 = 2GM/c^2$. Now suppose that s in eqn (1) is the length in a flat space with Cartesian coordinates w and r. Show that eqn 1 then defines a parabola

$$w^2 = 4r_0(r - r_0). \tag{2}$$

Now consider the equatorial plane in Schwarzschild space at constant time ($\theta = \pi/2$, $t =$ const.). Show that it is geometrically equivalent to a paraboloid of revolution obtained by rotating (2) around the w axis.

Q5.1 With the geometry of Fig. 5.1 show that the distance PP′ is given by

$$\begin{aligned} \mathrm{d}s^2 &= (\mathrm{d}x^1)^2 + (\mathrm{d}x^2)^2 + 2\,\mathrm{d}x^1\,\mathrm{d}x^2\cos\theta \\ &= \mathrm{d}x_1\,\mathrm{d}x^1 + \mathrm{d}x_2\,\mathrm{d}x^2. \end{aligned}$$

Q5.2 Show that $g^{\alpha\beta}g_{\alpha\beta} = n$ for an n-dimensional space.

Q5.3 Determine the transformation matrix Λ required to change from Cartesian coordinates (x, y) to polar coordinates (r, θ). Check that $\Lambda^{\alpha}{}_{\nu}\Lambda_{\beta}{}^{\nu} = \delta^{\alpha}{}_{\beta}$.

Q6.1 Show that for covector components the second term of eqn 6.8 becomes negative:

$$\frac{\mathrm{D}p_\mu}{\mathrm{D}x^\nu} = \frac{\partial p_\mu}{\partial x^\nu} - \Gamma^{\rho}_{\nu\mu}p_\rho.$$

(Take a scalar quantity $\varphi = p_\alpha q^\alpha$ and calculate its derivative.)

Q6.2 Show that a geodesic that is time-like (space-like, light-like) at a given place remains time-like (space-like, light-like) on its journey through space–time. Hint: consider parallel transporting a vector A_μ and show that $A_\mu A^\mu$ is constant.

Q6.3 Take a space-like geodesic in Minkowski space and show that it is neither maximal nor minimal.

Q7.1 Verify the expressions given for Γ^1_{00}, Γ^0_{10}, Γ^0_{10} and Γ^1_{11} for the Schwarzschild metric following eqn D.5 in Appendix D. Then calculate $\Gamma^1_{00,1}$, $\Gamma^1_{01,0}$. Finally calculate R^1_{010}.

Q7.2 Starting from the force given in eqn 7.21 show that the corresponding gravitational potential satisfies

$$\nabla^2\Phi = 4\pi G\rho - \Lambda c^2.$$

Q7.3 Show that the Riemann curvature tensor vanishes for the metric equation

$$\mathrm{d}s^2 = c^2\,\mathrm{d}t^2 - \mathrm{d}r^2 - r^2\,\mathrm{d}\theta^2.$$

Q8.1 What is the radar echo time delay for reflection from Mars at superior conjunction. Assume that the orbit of Mars is circular with a radius of 228 million km.

Q8.2 An observer is at rest at a distance r from a star of mass M. Show that his four-velocity in this frame (S) is $u^\alpha = (c/Z^{1/2}, 0, 0, 0)$ where

$Z = 1 - 2GM/rc^2$. Then consider a frame in free fall (F) which coincides momentarily with the observer's rest frame. Show that his velocity in the frame F is $v^\alpha = (c, 0, 0, 0)$. Suppose that a body of four-momentum p^α in frame S passes close by the observer. Show that the energy E^* of the body in frame F is $p^\alpha u_\alpha$.

This question illustrates that the energy measured by an observer is the invariant $p^\alpha u_\alpha$ where u^α is the observer's four-velocity and p^α is the body's four-momentum, both referred to the same frame.

Q9.1 Calculate the radius of the horizon of a neutral non-rotating black hole whose mass is $10^8\ M_\odot$. Show that the velocity of a body moving in the smallest stable orbit around this Schwarzschild black hole is $c/2$. Hence calculate the period as measured by a local observer at rest and that measured on a clock remote from the black hole (coordinate time).

Q9.2 If someone is in free fall radially towards a black hole which has a mass equal to $10\ M_\odot$ he would feel a lateral tidal force squeezing him. Calculate the radial distance from the centre of the black hole at which the tidal acceleration has grown to $400\ \mathrm{ms^{-2}\ m^{-1}}$. At this point our traveller will be crushed. We assume that the traveller starts from rest at a place remote from the black hole.

Q9.3 What is the Schwarzschild radius of a galactic black hole of $10^8\ M_\odot$? How much mass would need to flow into such a Schwarzschild black hole per year in order to generate a power of 10^{39} W?

Q10.1 Verify the expressions given in Section 10.3 for the xx and yy quadrupoles, and the reduced quadrupole moments of the system formed by the pulsar PSR 1913+16 and its companion.

Q10.2 Calculate the amplitude of the gravitational waves from the binary pulsar system PSR 1913+16 at the surface of the Earth.

Q10.3 Consider an aluminium gravitational antenna 5 m long weighing 10 t. What would be the range of gravitational wave frequencies to which it would respond? What would its sensitivity be if it is maintained at liquid helium temperature?

Q11.1 Consider three observers (A, B, C) who are remote from one another in the Universe. Show that if the motion of B and C relative to A is consistent with the Hubble expansion formula, then the motion of C relative to B is also consistent with this expression.

Q11.2 Using eqn 11.20 and the equation preceding it for a finite universe, show that the Universe has a maximum size R_c and a lifetime $\pi R_c/c$. Show that this maximum size is equal to the Schwarzschild radius of the Universe.

Q11.3 Derive the expression given for the component of the Ricci tensor in Section 11.3:

$$R_{11} = \frac{2k + R\ddot{R}/c^2 + 2\dot{R}^2/c^2}{1 - k\sigma^2}.$$

Q12.1 Show that the classical gravitational potential self-energy of a body of the Planck mass is equal to its rest mass energy.

Solutions

A1.1 (a) At velocity $0.8c$, $\beta = 0.8$ and $\gamma = 1.667$. Then the velocity four-vector is $(1.667, 1.333, 0, 0)c$, and the momentum four-vector is $(1563, 1251, 0, 0)\,\text{MeV}/c$.

(b) The velocity four-vector is $(1.667, 0.770, 0.770, 0.770)c$, and so the momentum four-vector is $(1563, 722, 722, 722)\,\text{MeV}/c$.

A1.2 Space-like; light-like (forward in time); light-like (forward in time); light-like (backward in time); space-like; time-like (forward in time).

A1.3 Zero: the neutrinos are massless and so their velocity is c; this makes the interval time-like.

A2.1 The gravitational red shift is $\Delta\nu/\nu = g\Delta r/c^2$ where Δr is the height of the tower (Section 2.2). When the values $g = 9.81\ \text{m s}^{-2}$, $\Delta r = 22.6\ \text{m}$ are inserted, $\Delta\nu/\nu = 2.46 \times 10^{-15}$.

A2.2 From eqn 2.1, $dt(r) = dt(\infty)(1 - GM/rc^2)$, and so $\nu(r) = \nu(\infty)(1 + GM/rc^2)$ and $\Delta\nu/\nu = GM/rc^2$. Now $\Delta\lambda/\lambda = \Delta\nu/\nu$ so that $\Delta\lambda = GM\lambda/rc^2$. For emission at the Sun's surface $\Delta\lambda = 1.63 \times 10^{-3}$ nm.

A3.1 The area of the 48 states is about $9 \times 10^6\ \text{km}^2$ and the curvature K of the Earth is $1/\text{radius}^2 = 2.46 \times 10^{-8}\ \text{km}^{-2}$, so that the rotation given by eqn 3.3 is 0.22 rad $\approx 12.7°$.

A3.2 Imagine the torus to be laid flat on the ground. Any plane vertical section containing the torus axis has two circular segments of radius of curvature 2 m. Now consider a horizontal section through the torus at its widest point. At the outer edge of the torus the section has radius $+12$ m and at the inner edge it has radius -8 m. Thus the torus surface at the outer edge has positive curvature $(1/2)(1/12) = +1/24 = 0.042\ \text{m}^{-2}$, and at the inner edge it has negative curvature $(1/2)(-1/8) = -0.062\ \text{m}^{-2}$. Close to the surface that rests in contact with the ground the curvature changes sign, i.e. becomes infinite.

A3.3 The radial term in the metric equation (Section 3.5) is $g_{rr} = 1/(1 + r^2/10^2)$ while $g_{\theta\theta} = r^2$ and $g_{\varphi\varphi}$ is $r^2 \sin^2\theta$. Using expression 3.7 the area is $4\pi \times 10^2 \sinh^2(1) = 1477\ \text{m}^2$ or about 17 per cent larger than in flat space.

A4.1 The position four-vector of the ball initially is (0, 0, 0, 0). As shown in Fig. I.1, at its highest point the ball reaches a height $\frac{1}{2}g(t/2)^2 = gt^2/8$. Its position four-vector is then $(ct/2, gt^2/8, 0, 0)$. Its final position four-vector is $(ct, 0, 0, 0)$. The interval in space–time along this

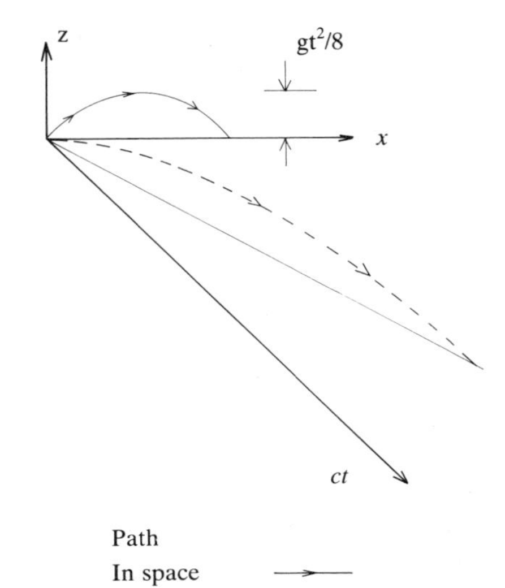

Fig. I.1

path is close to ct because $gt^2/8 \ll ct$. This path has a spatial sagitta $gt^2/8$. Then the radius of curvature is given by the sagitta formula

$$\rho = \frac{(ct/2)^2}{2(gt^2/8)} = \frac{c^2}{g} \approx 10^{16}\ \mathrm{m}$$

or approximately 1 light-year.

A4.2 The factor $2GM/rc^2$ in the metric equation evaluated at the surface is $2(6.6726 \times 10^{-11})(1.99 \times 10^{30})/(2.99792 \times 10^8)^2(3000) = 0.985$. The metric equation is $\mathrm{d}s^2 = 0.015\, c^2\, \mathrm{d}t^2 - 66.5\, \mathrm{d}r^2 - r^2\, \mathrm{d}\Omega$.

A4.3 In w–r space $\mathrm{d}s^2 = \mathrm{d}w^2 + \mathrm{d}r^2$ and so

$$\mathrm{d}w^2 = \frac{\mathrm{d}r^2}{1 - r_0/r} - \mathrm{d}r^2$$
$$= \frac{\mathrm{d}r^2}{r/r_0 - 1}.$$

Therefore

$$\mathrm{d}w = \frac{\mathrm{d}r}{(r/r_0 - 1)^{1/2}}$$

which on integrating gives

$$w = 2r_0\left(\frac{r}{r_0} - 1\right)^{1/2},$$

i.e.

$$w^2 = 4r_0(r - r_0). \tag{2}$$

The interval in the Schwarzschild equatorial plane is

$$ds^2 = \frac{dr^2}{1 - r_0/r} + r^2\, d\varphi^2.$$

If the parabola (2) is rotated about the w axis then we can call the angle of rotation φ, and the tangential distance along the surface generated is $r\, d\varphi$. Hence

$$ds^2 = r^2\, d\varphi^2.$$

For paths at constant φ we already have

$$ds^2 = \frac{dr^2}{1 - r_0/r}$$

and so distances across the surface when φ and r both change are given by

$$ds^2 = \frac{dr^2}{1 - r_0/r} + r^2\, d\varphi^2.$$

A5.1

$$(PP')^2 = (dx_1)^2 + (dx^2 \sin\theta)^2. \tag{1}$$

Now

$$dx_1 = dx^1 + dx^2 \cos\theta. \tag{2}$$

Therefore

$$(PP')^2 = (dx^1)^2 + (dx^2)^2 + 2\, dx^1\, dx^2 \cos\theta.$$

Also

$$dx_2 = dx^2 + dx^1 \cos\theta. \tag{3}$$

Forming the difference $\cos\theta(2) - (3)$ we obtain

$$dx_1 \cos\theta - dx_2 = dx^2(\cos^2\theta - 1).$$

Therefore

$$dx^2 = \frac{dx_2 - dx_1 \cos\theta}{\sin^2\theta}.$$

Substituting again in (1) for dx_1 and $dx^2 \sin^2\theta$, we obtain

$$(PP')^2 = dx_1(dx^1 + dx^2 \cos\theta) + dx^2(dx_2 - dx_1 \cos\theta)$$
$$= dx_1\, dx^1 + dx_2\, dx^2.$$

A5.2 If the metric is diagonal, eqn 5.13 shows that $g^{00} = 1/g_{00}$ etc. Then, choosing Cartesian coordinates locally,

$$g^{\alpha\beta}g_{\alpha\beta} = g^{00}g_{00} + g^{11}g_{11} + \cdots = n$$

for n dimensions (including the case that one dimension is temporal, and the other $n - 1$ dimensions are spatial). $g^{\alpha\beta}g_{\alpha\beta}$ is a scalar quantity, and so it has the same value independent of the frame, i.e. it is always equal to n.

A5.3 The relations between the coordinates are

$$r = (x^2 + y^2)^{1/2} \qquad \tan\theta = y/x.$$

Then

$$\Lambda^1{}_1 = \frac{dr}{dx} = \frac{x}{(x^2 + y^2)^{1/2}}$$

$$\Lambda^1{}_2 = \frac{dr}{dy} = \frac{y}{(x^2 + y^2)^{1/2}}.$$

Also

$$\sec^2\theta\, d\theta = \frac{x\, dy - y\, dx}{x^2}.$$

Therefore

$$d\theta = \frac{x\, dy - y\, dx}{x^2 + y^2}.$$

Hence

$$\Lambda^2{}_1 = -y/(x^2 + y^2)$$

$$\Lambda^2{}_2 = \frac{x}{x^2 + y^2}.$$

For the reverse operation

$$x = r\cos\theta \qquad y = r\sin\theta.$$

Then

$$\Lambda_1{}^1 = \cos\theta \qquad \Lambda_2{}^1 = -r\sin\theta$$

$$\Lambda_1{}^2 = \sin\theta \qquad \Lambda_2{}^2 = r\cos\theta.$$

For example,

$$\Lambda^1{}_1\Lambda_1{}^1 + \Lambda^1{}_2\Lambda_1{}^2 = \frac{x\cos\theta + y\sin\theta}{r} = 1$$

$$\Lambda^1{}_1\Lambda_2{}^1 + \Lambda^1{}_2\Lambda_2{}^2 = -x\sin\theta + y\cos\theta = 0.$$

A6.1

$$\frac{\partial\varphi}{\partial x^\nu} = \frac{\partial p_\mu}{\partial x^\nu}q^\mu + p_\mu\frac{\partial q^\mu}{\partial x^\nu}$$

$$= \frac{\partial p_\mu}{\partial x^\nu}q^\mu + p_\mu\frac{\mathrm{D}q^\mu}{\mathrm{D}x^\nu} - p_\mu q^\alpha\Gamma^\mu_{\nu\alpha}$$

$$= \frac{\partial p_\mu}{\partial x^\nu}q^\mu + p_\mu\frac{\mathrm{D}q^\mu}{\mathrm{D}x^\nu} - p_\rho q^\mu\Gamma^\rho_{\nu\mu}.$$

Therefore

$$\frac{\partial\varphi}{\partial x^\nu} = q^\mu\left(\frac{\partial p_\mu}{\partial x^\nu} - p_\rho\Gamma^\rho_{\nu\mu}\right) + p_\mu\frac{\mathrm{D}q^\mu}{\mathrm{D}x^\nu}. \quad (1)$$

Now because φ is a scalar quantity

$$\frac{\partial\varphi}{\partial x^\nu} \equiv \frac{\mathrm{D}\varphi}{\mathrm{D}x^\nu}.$$

Therefore

$$\frac{\partial\varphi}{\partial x^\nu} = q^\mu\frac{\mathrm{D}p_\mu}{\mathrm{D}x^\nu} + p_\mu\frac{\mathrm{D}q^\mu}{\mathrm{D}x^\nu} \quad (2)$$

Comparing the two expressions for $\partial\varphi/\partial x^\nu$ we have

$$\frac{\mathrm{D}p_\mu}{\mathrm{D}x^\nu} = \frac{\partial p_\mu}{\partial x^\nu} - p_\rho\Gamma^\rho_{\nu\mu}.$$

A6.2 If A_μ is parallel transported, then $\mathrm{D}A^\mu/\mathrm{D}\lambda = 0$, and

$$\frac{\mathrm{D}(A^\mu A_\mu)}{\mathrm{D}\lambda} = \frac{A^\mu\,\mathrm{D}A_\mu}{\mathrm{D}\lambda} + \frac{A_\mu\,\mathrm{D}A^\mu}{\mathrm{D}\lambda} = 0.$$

Along a geodesic the path element $\mathrm{d}x^\mu$ itself is being parallel transported and so

$$\frac{\mathrm{D}(\mathrm{d}x^\mu/\mathrm{d}\lambda)}{\mathrm{D}\lambda} = 0.$$

Thus $(\mathrm{d}x^\mu/\mathrm{d}\lambda)(\mathrm{d}x_\mu/\mathrm{d}\lambda)$ remains constant. Hence $\mathrm{d}x^\mu\,\mathrm{d}x_\mu$ continues time-like (or light- or space-like) if it started that way.

A6.3 Let the end-points be at $A(t = 0, x = 0)$ and $\mathrm{C}(t = 0, x = x)$. Consider the path ABC through $\mathrm{B}(t = 0, x = 0, y = y)$. The various paths elements have intervals

$$S^2_{\mathrm{AC}} = -x^2, \qquad S^2_{\mathrm{AB}} = -y^2, \qquad S^2_{\mathrm{BC}} = -x^2 - y^2.$$

Then

$$S_{AC}^2 > S_{AB}^2 + S_{BC}^2.$$

An alternative path lies through D($t = t$, $x = 0$). The new intervals are

$$S_{AD}^2 = ct^2 \qquad S_{DC}^2 = ct^2 - x^2.$$

Then

$$S_{AC}^2 < S_{AD}^2 + S_{DC}^2.$$

Therefore AC is neither maximal nor minimal.

A7.1(a)

$$2\Gamma^1_{00} = 2g^{11}\Gamma_{100} = -Z(g_{10,0} - g_{00,1} + g_{10,0})$$
$$= Zg_{00,1} = Z(m/r^2),$$

Therefore

$$\Gamma^1_{00} = mZ/2r^2.$$

$$2\Gamma^0_{01} = 2g^{00}\Gamma_{001} = (1/Z)(g_{00,1} - g_{01,0} + g_{01,0})$$
$$= (1/Z)g_{00,1} = (1/Z)(m/r^2).$$

Therefore

$$\Gamma^0_{01} = m/2r^2Z.$$

Now

$$\Gamma^0_{10} = g^{00}\Gamma_{010} = g^{00}\Gamma_{001}.$$

Therefore

$$\Gamma^0_{10} = m/2r^2Z.$$

$$2\Gamma^1_{11} = 2g^{11}\Gamma_{111} = -Z(g_{11,1} - g_{11,1} + g_{11,1})$$
$$= -Z(g_{11,1}) = -Z\left(\frac{1}{Z^2}\right)\left(\frac{m}{r^2}\right).$$

Therefore

$$\Gamma^1_{11} = \frac{-m}{2r^2Z}.$$

(b)

$$\Gamma^1_{00,1} = \frac{m(\mathrm{d}Z/\mathrm{d}r)}{2r^2} - mZ/r^3 = m^2/2r^4 - mZ/r^3.$$

Γ^1_{01} does not depend on time, and so $\Gamma^1_{01,0} = 0$.

(c)

$$R^1_{010} = \Gamma^1_{00,1} - \Gamma^1_{01,0} + \Gamma^1_{\sigma 1}\Gamma^\sigma_{00} - \Gamma^1_{\sigma 0}\Gamma^\sigma_{01}$$
$$= \Gamma^1_{00,1} + \Gamma^1_{11}\Gamma^1_{00} - \Gamma^1_{00}\Gamma^0_{01}$$
$$= m^2/2r^4 - mZ/r^3 - m^2/4r^4 - m^2/4r^4,$$

i.e.

$$R^1_{010} = -mZ/r^3.$$

A7.2 The flux of the force field out of a sphere of radius r containing a mass M is

$$\int \mathbf{F}\cdot d\mathbf{s} = -4\pi GM + 4\pi c^2\Lambda r^3/3$$

$$= -4\pi G \int \rho \, dV + c^2\Lambda \int dV,$$

where the volume intergrals are performed over the same sphere. Applying Stokes's theorem, we obtain

$$\int \nabla\cdot\mathbf{F}\, dV = -4\pi G \int \rho\, dV + c^2\Lambda \int dV.$$

Because this holds for any volume we have

$$\nabla\cdot\mathbf{F} = -4\pi G\rho + c^2\Lambda$$

so that

$$\nabla^2\Phi = 4\pi G\rho - c^2\Lambda.$$

A7.3 Taking results from Chapter 6,

$$g_{00} = g^{00} = 1, \qquad g_{11} = g^{11} = -1, \qquad g_{22} = \frac{1}{g^{22}} = -r^2.$$

Then the only non-vanishing derivative is $g_{22,1} = -2r$. Using eqn 6.5 we obtain

$$\Gamma^2_{12} = \Gamma^2_{21} = 1/r \qquad \text{and} \qquad \Gamma^1_{22} = -r.$$

Also, $\Gamma^2_{12,1} = \Gamma^2_{21,1} = -1/r^2$ and $\Gamma^1_{22,1} = -1$. Finally using eqn 7.3,

$$R^1_{212} = \Gamma^1_{22,1} - \Gamma^1_{22}\Gamma^2_{21} = 0.$$

A8.1 The time delay is given by eqn 8.17, with r_V replaced by the radius r_M of the orbit of Mars round the Sun. Then

$$\Delta t = \frac{4GM}{c^3}\left[\ln\left(\frac{4r_E r_M}{b^2}\right) + 1\right]$$

where M is the Sun's mass and b is the Sun's radius. Therefore

$$\Delta t = 1.98 \times 10^{-5}[\ln(2.82 \times 10^5) + 1] = 270\ \mu\text{s}.$$

A8.2 The observer's velocity satisfies the relation

$$g_{\alpha\beta}u^\alpha u^\beta = c^2$$

in all frames. We choose spherical polar coordinates. He is at rest in frame S and there

$$u^1 = u^2 = u^3 = 0.$$

The above equation gives

$$g_{00}u^0u^0 = Zu^0u^0 = c^2,$$

so that

$$u^0 = c/Z^{1/2}.$$

In the frame F (freely falling) we again have

$$v^1 = v^2 = v^3 = 0$$

while now

$$g_{00}v^0v^0 = v^0v^0 = c^2.$$

Thus

$$v^0 = v_0 = c.$$

If q^α is the four-momentum of the body in frame F, then

$$p^\alpha u_\alpha = q^\alpha v_\alpha = cq^0 = E^*.$$

Evaluating this in frame S gives

$$p^\alpha u_\alpha = g_{00}p^0u^0 = cp^0Z^{1/2}.$$

Thus

$$E^* = cp^0Z^{1/2} = \mathrm{c}p_0/Z^{1/2}.$$

A9.1 The radius of the horizon is $2GM/c^2 = 2.96 \times 10^{11}$ m. Using equation 9.13 the radius is 8.89×10^{11} m. The orbital velocity is obtained from eqn 9.14. Rewritten using eqn 8.3, this becomes

$$Zmc^2\left(\frac{\mathrm{d}t}{\mathrm{d}\tau}\right) = \left(\frac{8}{9}\right)^{1/2} mc^2$$

with

$$Z = 1 - r_0/r = 2/3.$$

Thus

$$\frac{\mathrm{d}t}{\mathrm{d}\tau} = 2^{1/2} \qquad \text{and} \qquad Z\left(\frac{\mathrm{d}t}{\mathrm{d}\tau}\right)^2 = \frac{4}{3}.$$

If v_φ is the tangential velocity we have

$$c^2 = 4c^2/3 - v_\varphi^2.$$

It follows that

$$v_\varphi = \beta\gamma = 1/3^{1/2} \qquad \text{and} \qquad \beta = 1/2.$$

Thus the period as measured by a stationary local observer is

$$\tau = 2\pi r/\beta c = 3.74 \times 10^4 \text{ s}.$$

The coordinate time taken is $\tau/Z^{1/2} = 4.5 \times 10^4$ s.

A9.2 The relative acceleration of parts of the traveller are determined by the equation of geodesic deviation 7.6. We are interested in the tangential component

$$\frac{D^2\xi^2}{D\tau^2} = -R^2{}_{\nu 2\lambda}\xi^2 \frac{dx^\nu}{d\tau}\frac{dx^\lambda}{d\tau}.$$

Values of the components of the Riemann tensor can be found in Appendix D. Only R^2_{020} and R^2_{121} are required because the motion is radial. Then

$$Q = \frac{D^2\xi^2/D\tau^2}{\xi^2} = -\frac{r_0 Z}{2r^3}\left(\frac{c\,dt}{d\tau}\right)^2 + \frac{r_0}{2Zr^3}\left(\frac{dr}{d\tau}\right)^2$$

where $r_0 = 2GM/c^2$ and $Z = 1 - r_0/r$. Equations 8.3 and 8.6 determine $dt/d\tau$ and $dr/d\tau$ in terms of Z. Because the person starts at rest a long way from the black hole we have $E = mc^2$. Thus eqn 8.3 becomes

$$\frac{dt}{d\tau} = \frac{1}{Z},$$

while eqn 8.6 becomes

$$\left(\frac{dr}{d\tau}\right)^2 = \frac{2GM}{r}.$$

Then

$$Q = -\frac{r_0 c^2}{2Zr^3}\left(1 - \frac{2GM}{rc^2}\right) = -\frac{r_0 c^2}{2r^3},$$

i.e.

$$Q = -GM/r^3.$$

This reproduces eqn 4.5, which we could also use as our starting point. When Q reaches 400 m s^{-2} m^{-1},

$$r = 1.49 \times 10^6 \text{ m}.$$

A9.3 The Schwarzschild radius of a mass of $10^8\ M_\odot$ is $r_0 = 2G(10^8\ M_\odot)/c^2 = 3 \times 10^{11}$ m, or about 1000 light-seconds. If the material falls into the minimum circular stable orbit, the energy released per unit mass is $0.0572c^2$, or 1.03×10^{46} J/$M_\odot$. The desired power output therefore requires $10^{-7}\ M_\odot$ per second or $3.1\ M_\odot$ per year.

A10.1 The orbits are drawn in Fig. I.2 with the two masses M circling the centre of mass O at a radius a. At time t, as drawn, the contribution of the pulsar to I_{xx} is $Mx^2 = M(a \cos \omega t)^2$. The companion makes an equal contribution. Thus

$$I_{xx} = 2Ma^2 \cos^2 \omega t = Ma^2(1 + \cos 2\omega t).$$

Similarly

$$I_{yy} = 2My^2 = 2Ma^2 \sin^2 \omega t = Ma^2(1 - \cos 2\omega t).$$

For the cross-terms

$$I_{xy} = I_{yx} = 2Ma^2 \cos \omega t \sin \omega t = Ma^2 \sin 2\omega t.$$

There is no motion in the z direction, and so we can work in two dimensions. Then the expression for the traceless or reduced quadrupole moment becomes

$$\not{I}_{ij} = \int (x_i x_j - \delta_{ij} x_k{}^2/2)\rho \, \mathrm{d}V$$

where i, j, and k run from 1 to 2 only. In the present example

$$\not{I}_{xx} = I_{xx} - Ma^2 = Ma^2 \cos 2\omega t$$

$$\not{I}_{yy} = I_{yy} - Ma^2 = -Ma^2 \cos 2\omega t$$

$$\not{I}_{xy} = I_{xy} = Ma^2 \sin 2\omega t$$

$$\not{I}_{yx} = \not{I}_{xy}.$$

A10.2 From Section 10.3 we see that the luminosity of the source is

$$L = (G/5c^5)(128\omega^6 M^2 a^4),$$

where ω is the orbital angular frequency (2.25×10^{-4} rad s^{-1}), M is

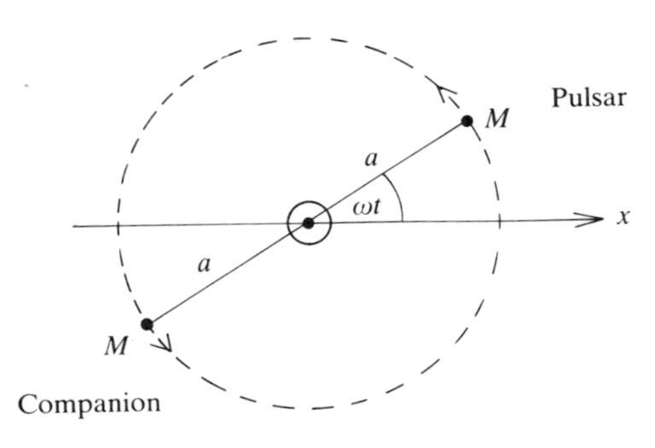

Fig. I.2

the mass of each star (2.82×10^{30} kg), and a is the orbital radius (3.08 light-seconds). The energy flux at the Earth is

$$F = L/4\pi r^2,$$

where r is the distance of the pulsar from the Earth (1.54×10^{20} m). Also, from eqn E.9 we have

$$F = \frac{c^3}{32\pi G}\langle \dot{h}_{ij}{}^2 \rangle = \frac{c^3\omega^2}{32\pi G}\langle h_{ij}{}^2 \rangle,$$

whence

$$\langle h_{ij}{}^2 \rangle = \frac{32\pi GF}{c^3\omega^2} = \frac{8GL}{c^3\omega^2 r^2} = \frac{1024 G^2\omega^4 M^2 a^4}{5c^8 r^2},$$

i.e.

$$h_{ij} \approx \frac{14.3 G\omega^2 M a^2}{c^4 r} = 9.3 \times 10^{-23},$$

with of course a period of 7.75 h.

A10.3 The natural frequency of longitudinal oscillation $\nu_0 = v_s/2l$ where v_s is the velocity of sound in the bar and l is its length. In the present case $\nu_0 = 550$ Hz. If the Q value were 10^6 then the frequency range would be about $550/10^6 \approx 0.5$ mHz. The quantum limit on the detectable strain is (Section 10.5)

$$h(\text{min}) = \left(\frac{\hbar}{2\pi l^2 \nu_0 M}\right)^{1/2},$$

where M is the mass of the bar. This yields

$$h(\text{min}) = 3.5 \times 10^{-22}.$$

A11.1 If B appears to be receding from A at a velocity given by the Hubble formula,

$$v_{\text{BA}} = z_{\text{BA}} c = H d_{\text{BA}}.$$

Because the motion appears to be radially away from A we can write this vectorially:

$$\boldsymbol{v}_{\text{BA}} = H \boldsymbol{d}_{\text{BA}}.$$

Similarly

$$\boldsymbol{v}_{\mathrm{CA}} = H\boldsymbol{d}_{\mathrm{CA}}.$$

Taking the difference between the last two equations gives

$$\boldsymbol{v}_{\mathrm{CA}} - \boldsymbol{v}_{\mathrm{BA}} = H(\boldsymbol{d}_{\mathrm{CA}} - \boldsymbol{d}_{\mathrm{BA}}),$$

i.e.

$$\boldsymbol{v}_{\mathrm{CB}} = H\boldsymbol{d}_{\mathrm{CB}}.$$

Thus the relative motion of C with respect to B is consistent with Hubble's law. The type of motion described by Hubble's law is therefore in accord with the cosmological principle.

A11.2 Using the relation $c\,\mathrm{d}t = \mathrm{d}R/(1/\mu - 1)^{1/2}$ we see that when the Universe reaches its maximum size, and hence $\mathrm{d}R/\mathrm{d}t = 0$, then $\mu = 1$. At this stage the parameter

$$R = R_{\mathrm{C}} = \frac{8\pi G\rho_0 R_0{}^3}{3c^2}.$$

If M_0 is the mass of matter in the Universe, then

$$R = R_{\mathrm{C}} = 2GM_0/c^2.$$

Now use eqn 11.20 to give the time taken for the Universe to reach this maximum size:

$$t = \pi R_{\mathrm{C}}/2c.$$

The total lifetime of the Universe is twice this time because the collapse is the mirror image of the expansion; thus the total lifetime is

$$2t = \pi R_{\mathrm{C}}/c$$

A11.3 The expression for this component is

$$R_{11} = R^0{}_{101} + R^1{}_{111} + R^2{}_{121} + R^3{}_{131},$$

of which $R^1{}_{111}$ is shown to be identically zero in Appendix B. Evaluating the remaining terms in order, we have

$$R^0{}_{101} = g^{00}R_{0101} = g^{00}R_{1010} = g_{11}g^{00}R^1{}_{010}$$
$$= \frac{R\ddot{R}}{c^2(1 - k\sigma^2)}.$$

We also have from eqn 6.6 that

$$\Gamma^0{}_{11} = g^{00}\Gamma_{011} = -g^{00}g_{11,0}/2 = R\dot{R}/c(1 - k\sigma^2)$$
$$\Gamma^2{}_{02} = \Gamma^2{}_{20} = g^{22}\Gamma_{220} = g^{22}g_{22,0}/2 = \dot{R}/Rc$$
$$\Gamma^2{}_{12} = \Gamma^2{}_{21} = g^{22}\Gamma_{221} = g^{22}g_{22,1}/2 = 1/\sigma$$
$$\Gamma^3{}_{13} = \Gamma^3{}_{31} = g^{33}\Gamma_{331} = g^{33}g_{33,1}/2 = 1/\sigma$$

$$\Gamma^1{}_{11} = g^{11}g_{11,1}/2 = k\sigma/(1 - k\sigma^2)$$
$$\Gamma^1{}_{01} = \Gamma^1{}_{10} = g^{11}\Gamma_{110} = g^{11}g_{11,0}/2 = \dot{R}/Rc$$
$$\Gamma^3{}_{03} = \Gamma^3{}_{30} = g^{33}\Gamma_{330} = g^{33}g_{33,0}/2 = \dot{R}/Rc.$$

Thus

$$\Gamma^2{}_{12,1} = \Gamma^3{}_{13,1} = -1/\sigma^2.$$

Then, applying eqn 7.3, we have

$$\begin{aligned} R^2{}_{121} &= \Gamma^2{}_{11,2} - \Gamma^2{}_{12,1} + \Gamma^2{}_{\sigma 2}\Gamma^\sigma{}_{11} - \Gamma^2{}_{\sigma 1}\Gamma^\sigma{}_{12} \\ &= -\Gamma^2{}_{12,1} + \Gamma^2{}_{02}\Gamma^0{}_{11} + \Gamma^2{}_{12}\Gamma^1{}_{11} - \Gamma^2{}_{21}\Gamma^2{}_{12}, \end{aligned}$$

leaving out terms that vanish. Therefore

$$\begin{aligned} R^2{}_{121} &= \frac{1}{\sigma^2} + \frac{\dot{R}^2}{c^2(1 - k\sigma^2)} + \frac{k}{1 - k\sigma^2} - \frac{1}{\sigma^2} \\ &= \frac{\dot{R}^2/c^2 + k}{1 - k\sigma^2}. \end{aligned}$$

Similarly from eqn 7.3,

$$\begin{aligned} R^3{}_{131} &= \Gamma^3{}_{11,3} - \Gamma^3{}_{13,1} + \Gamma^3{}_{\sigma 3}\Gamma^\sigma{}_{11} - \Gamma^3{}_{\sigma 1}\Gamma^\sigma{}_{13} \\ &= -\Gamma^3{}_{13,1} + \Gamma^3{}_{03}\Gamma^0{}_{11} + \Gamma^3{}_{13}\Gamma^1{}_{11} - \Gamma^3{}_{31}\Gamma^3{}_{13} \\ &= \frac{1}{\sigma^2} + \frac{\dot{R}^2}{c^2(1 - k\sigma^2)} + \frac{k}{(1 - k\sigma^2)} - \frac{1}{\sigma^2} \\ &= \frac{\dot{R}^2/c^2 + k}{1 - k\sigma^2}. \end{aligned}$$

Collecting terms, we have finally

$$R_{11} = \frac{R\ddot{R}/c^2 + 2\dot{R}^2/c^2 + 2k}{1 - k\sigma^2}.$$

A12.1 The gravitational potential self-energy of a mass M_p is approximately $GM_p{}^2/R_p$. The rest energy of the same mass is M_pc^2. Now

$$\frac{GM_p{}^2}{R_p} = \frac{G(\hbar c/G)}{(\hbar G/c^3)^{1/2}} = \left(\frac{c^5\hbar}{G}\right)^{1/2}$$

and

$$M_pc^2 = \left(\frac{\hbar c}{G}\right)^{1/2} c^2 = \left(\frac{c^5\hbar}{G}\right)^{1/2},$$

which proves the assertion.

References

Amaldi, E., Pizzella G., *et al.* (1986). *Il Nuovo Cimento* **9**, 829–44.

Bernstein, J., Brown, L. S., and Feinberg, G. (1989). *Review of Modern Physics* **61**, 25–39.

Boesgaard and Steigman (1985). *Annual Reviews of Astronomy and Astrophysics* **23**, 319–49.

Chandrasekhar, S. (1931). *Astrophysical Journal* **74**, 81–91.

Claverie, A., Isaak, G. R., McLeod, C. P., Van der Raay, H. B. and Roca Cortes, T. (1981). *Nature* **293**, 443–5.

Davis, M., Ruffini, R., Press, W. H., and Price, R. H. (1971). *Physical Review Letters*, **27**, 1466–9.

Dicke, R. H., Kuhn, J. R., and Libbrecht, K. G. (1985). *Nature* **316**, 687–90.

Eddington, A. S. (1924). *Nature, London* **113**, 192–6.

Felten, J. E. and Isaacman, R. (1986). *Review of Modern Physics*, **58**, 689–98.

Finkelstein, D. (1958). *Physical Review*, **110**, 965–7.

Fomalont, E. B. and Sramek, R. A. (1976). *Physical Review Letters* **36**, 1475–8.

Friedman, A. (1922). *Zeitschrift für Physik* **10**, 377.

Friedman, J. L., Isper, J. R., and Parker, L. (1984). *Nature, London* **312**, 255–7.

Guth, A. H. and Steinhardt, P. J. (1984). *Scientific American* May, 90–102.

Hawking, S. W. (1974). *Communications in Mathematical Physics* **42**, 199–200.

Hawking, S. W. (1974). *Nature, London* **248**, 30–1.

Hellings, R. W. (1984). In *General relativity and gravitation* (eds B. Bertotti, F. de Felice, and A. Pascolini), pp. 365–85. Reidel, Dordrecht.

Hubble, E. P. (1929). *Proceedings of the National Academy of Sciences of the United States of America* **15**, 169.

Hubble, E. P. and Humason, M. L. (1931). *Astrophysical Journal* **74**, 43–50.

Hulse, R. A. and Taylor, J. H. (1975). *Astrophysical Journal* **195**, L51–3.

Kaluza, Th. (1921). *Sitzungberichte der K. Prussischen Akademie der Wissen zu Berlin* K1, 966–74.

Klein, O. (1926). *Zeitschrift für Physik* **37**, 895–900.

Koester, L. (1976). *Physical Review D* **14**, 907–9.

Kruskal, M. D. (1960). *Physical Review* **119**, 1743–4.

Kuroda, K. and Mio, N. (1989). *Physical Review Letters* **62**, 1941–4.

Le Maître, G. (1927). *Annales de la Societé Scientifique de Bruxelles, A* **47**, 49.

Linde, A. D. (1987). *Physics Today* September, 61–8.

Lockhart, J. M., Witteborn, F. C., and Fairbank, W. M. (1977). *Physical Review Letters* **38**, 1220–3.

Matsumoto, T. *et al.* (1988). *Astrophysical Journal* **329**, 567–71.

McClintock, J. (1988). *Sky and Telescope* January, 28–33.

Michell, J. (1784). *Philosophical Transactions of the Royal Society of London* **74**, 35.

Misner, C. W., Thorne, K. S., and Wheeler, J. A. (1972). *Gravitation.* Freeman, San Francisco, CA.

Ninkov, Z., Walker, G. A. H., and Yang, S. (1987). *Astrophysical Journal* **321**, 425–37.
Oppenheimer, J. R. and Volkoff, G. M. (1939). *Physical Review* **55**, 374–8.
Peebles, P. J. E. (1971). *Physical cosmology*. Princeton University Press, Princeton, NJ.
Peebles, P. J. E. (1986). *Nature, London* **321**, 27–32.
Penrose, R. (1969). *Rivista Nuovo Cimento* **1**, 252–76.
Penzias, A. A. and Wilson, R. W. (1965). *Astrophysical Journal* **142**, 419.
Pound, R. V. and Rebka, G. A. (1960). *Physical Review Letters* **4**, 337–41.
Press and Thorne (1972). *Annual Reviews of Astronomy and Astrophysics* **10**, 335–74.
Reasenberg, R. D. *et al.* (1979). *Astrophysical Journal* **234**, L219–21.
Robertson, H. P. (1936). *Astrophysical Journal* **82**, 257.
Roll, P. G., Krotkov, R., and Dicke, R. H. (1964). *Annals of Physics* (*New York*) **26**, 442–517.
Rubin, V. C. and Ford, W. K. Jr. (1980). *Astrophysical Journal* **238**, 471.
Schwarz, J. H. (1984). *Comments on nuclear and particle Physics* **13**, 103–8.
Schwarz, J. H. (1985). *Comments on nuclear and particle Physics* **15**, 9–15.
Shapiro, I. I. *et al.* (1971). *Physical Review Letters* **26**, 1132–5.
Shapiro, I. I. *et al.* (1972). *Physical Review Letters* **28**, 1594–7.
Shapiro, S. L. and Teukolsky. S. A. (1983). *Black holes, white dwarfs and neutron stars.* Wiley, New York.
Silk, J. (1980). *The big bang, the creation and evolution of the Universe*, Freeman, San Francisco, CA.
Stella, L. *et al.* (1985). *Astrophysical Journal* **288**, L45–9.
Surdej, J. *et al.* (1987). *Nature, London* **329**, 695–6.
Taylor, J. H. (1986). In *General relativity and gravitation* (ed. M. A. H. MacCallum), pp. 209–22. Cambridge University Press, Cambridge.
Vessot, R. F. C. *et al.* (1980). *Physical Review Letters* **45**, 2081–4.
Walker, A. G. (1936). *Proceedings of the London Mathematical Society* **42**, 90.
Weber, J. (1960). *Physical Review* **117**, 306–13.
Weissberg, J. M. and Taylor, J. H. (1984). *Physical Review Letters* **52**, 1348–50.
Wesson, P. S., Valle, K., and Stabell, R. (1987). *Astrophysical Journal* **317**, 601–6.
Will, C. M. (1981) Theory and experiment in gravitational physics. Cambridge University Press, Cambridge.
Witteborn, F. C. and Fairbank, W. M. (1968). *Nature, London* **220**, 436.

General bibliography

Einstein, A., *The meaning of relativity*, Chapman and Hall, London, 5th edn (1951).
This is a clear and compact account of the subject by its originator. It is based on a translation made in 1921 and therefore has little to say about modern developments.

Misner, C. W., Thorne, K. S., and Wheeler, J. A., *Gravitation*, Freeman, San Francisco, CA (1973).
A monolithic text covering general relativity, gravitation, and related aspects of cosmology. Parallel treatments at two levels are given: one is simple and physical, while the other contains a deeper and more detailed analysis. Many interesting points are brought out using a wealth of illustrations.

Weinberg, S., *Gravitation and Cosmology*, Wiley, New York (1973).
A thorough and elegant account by an eminent theoretical physicist.

Hawking, S. W. and Israel, W. (eds), 300 *years of gravitation*, Cambridge University Press, Cambridge (1987); *General relativity: an Einstein centenary survey*, Cambridge University Press, Cambridge (1979).
These two volumes contain the proceedings of conferences held to mark important anniversaries. They include surveys by experts on the current status of studies in gravitation and general relativity.

Chapter 1

French, A. P., *Special relativity*, Norton, New York (1968).
An excellent introductory text for special relativity.

Chapter 2

Rindler, W., *Essential relativity*, Springer-Verlag, Berlin, 2nd edn (1977).
A presentation of special and general relativity from a physical viewpoint.

French, A. P., *Newtonian mechanics*, Norton, New York (1976).
A basic text giving good coverage of the calculations of planetary orbits, tidal forces, and the centrifugal and Coriolis forces.

Chapters 3 and 4

Rindler, W., *Essential relativity*.

Laugwitz, D., *Differential and Riemannian geometry*, Academic Press, New York (1965).
A thorough review of these subjects.

Struick, D. J., *Lectures in classical differential geometry*, Dover Publications, Mineola, NY, 2nd edn (1961).
An excellent introduction to the concepts of curvature and curved spaces.

Chapters 5, 6 and 7

Laugwitz, D., *Differential and Riemannian geometry*.
Schutz, B. F., *A first course in general relativity*, Cambridge University Press, Cambridge (1985).
Lawdon, D. F., *An introduction to tensor calculus, relativity and cosmology*, Wiley, New York, 3rd edn (1982).

Chapter 8

French, A. P., *Newtonian mechanics*.
Will, C. M., *Was Einstein right*?, Basic Books, New York (1986).
A readable account of tests of Einstein's theory, backed by expert knowledge of the subject.
Will, C. M., *Theory and experiment in gravitational physics*, Cambridge University Press, Cambridge (1981).
In-depth presentation of the analysis of modern experiments. The post-Newtonian parametrization is used to provide a general framework within which to discuss the relative performance of general relativity and alternative theories.

Chapter 9

Shapiro, S. L. and Teukolsky, S. A., *Black holes, white dwarfs and neutron stars*, Wiley, New York (1983).
An excellent and authoritative account of the physics of compact astronomical objects.

Chapter 10

Blair D. C. (1987). *Contemporary Physics* **28**, 457–75.
Schutz, B. F., *A first course in general relativity*.
Weinberg, S., *Gravitation and cosmology*.

Chapter 11

Berry, M., *Principles of cosmology and gravitation*, Cambridge University Press, Cambridge (1976).
A clear and compact introduction to cosmology and gravitation. It contains a survey of the methods of determining masses and distances on the cosmic scale.
Bethe, H. A. and Brown, G. (1985). *Scientific American* **252**, 40–8.
Kippenhahn, R., *One hundred billion suns*, Unwin Paperbacks, London (1983).
An excellent readable account of stellar processes and cosmology.
Sciama, D. W., *Cosmology*, Cambridge University Press, Cambridge (1971).
A basic undergraduate text on the subject of cosmology.
Silk, J., *The big bang, the creation and evolution of the universe*, Freeman, San Francisco, CA (1980).
A highly readable account of modern cosmology and the origins of the Universe.

Peebles, P. J. E., *Physical cosmology*, Princeton University Press, Princeton, NJ (1971).
A detailed analysis of the cosmic background radiation, recombination and related matters.

Chapter 12

Narlikar, J. V. and Padmanabhan, T., *Gravity, gauge theories and quantum cosmology*, Reidel, Dordrecht (1986).
An up-to-date account of the topics broached in chapter 12.
Hawking, S. W. and Ellis, G. F. R., *The large scale structure of space–time*, Cambridge University Press, Cambridge (1973).
An advanced text which offers a deeper analysis of matters such as singularities in space–time.

Index

acceleration 9, 46, 68–70, 73, 86, 130
affine connections, *see* metric connections
age of universe 147, 160–1

Bianchi identities 190
big bang model 147, 152, 160, 163, 170
binary pulsar
  general relativistic effects in 100–5
  gravitational radiation from 104–5, 134–6
Birkhoff's theorem 44
black hole
  angular momentum of 115–16
  formation of 118–21
  Hawking radiation from 116–18
  Kerr 115
  lifetime of 117
  Schwarzschild 106
  stable orbits around a 113–15
  temperature of 117

centrifugal force 9, 69–70
Chandrasekhar, S. 2, 118
  limit 118–20
Christoffel symbols, *see* metric connections
closed universe 161
comoving coordinates 149
conservation
  of angular momentum 90
  of energy-momentum 6, 89–90
coordinate time 17, 44
Coriolis force 69–70
cosmic background radiation (CBR) 3, 144–6, 151–4, 163–5
  anisotropy of 145
  formation of 164–5
  last scattering of 165
  temperature of 3, 145
cosmological constant 82–3, 85, 163, 171–2
cosmological principle 147, 154, 222
covariant derivative 58–63
  for contravariant components 58–61
  for covariant components 62
covector component 49–52
curvature
  and tidal force 41–2, 73, 86
  Gaussian, *see* Gaussian curvature
  space–time with constant 147–54
  space with constant 35–7
Cygnus X-1 121–3

dark matter 158
deceleration parameter 151
decoupling era 165
deflection of light 1, 2, 19–20, 93–5, 98–9
density of matter in universe
  critical 158
  visible 156–8
density of radiation in universe 163–4
derivative, *see* convariant derivative
divergence
  of stress-energy tensor 79–81
  of Einstein tensor 82, 191
Doppler shift 14, 18, 100, 122–3
dust filled universe 157
dynamics for models of universe 154–7

earth
  precession of orbit 93
Eddington, A.S. 1
  coordinates 111
Einstein, A. 1, 4, 9–11, 15, 19 40, 57, 63, 72, 82–3, 87, 98
Einstein-de Sitter Universe 159
Einstein equation 82
  for Robertson-Walker model 154–7
  in Newtonian limit 84–6
  in vacuum 83
Einstein summation convention 39
Einstein tensor
  definition of 82
  properties of 82, 191
  in Robertson-Walker model 155
  on a spherical surface 193
electric charge
  of a black hole 115
element abundances 147, 165
elliptical orbit 91–3
energy
  conservation 80–1
  gravitational energy release 114–15, 124, 136–8
  in gravitational waves 133, 137–8, 198–200
  of orbit in Schwarzschild metric 89–90, 113–15
  release and inflation in early universe 171–2
energy-momentum four-vector 6–7, 77

energy-momentum tensor, *see* stress-energy tensor
equation of state
  for matter-dominated universe 157–8, 160–1
  for radiation-dominated universe 163–4
equivalence principle
  strong 11–12, 15–20, 40, 57, 94
  weak 9–10, 12–15, 179–80
ergosphere 115
Euclidean spaces 22, 24, 26, 39–40, 51
event in space–time 6–8
expansion of universe 146–7, 151–4, 157–72
  and red-shift 146, 151

Feynman diagrams 177–9
fifth force 14–15
flat space 21–23, 38
force 63
  covariant components 49–50
  four vector definition 7–8
four momentum, *see* energy-momentum
four vector 6–8
four velocity 7, 79, 121, 154
Friedmann's equations 156–7

galaxy, size of our 144
Galileo 9
gaue theory 176–9
Gaussian coordinates 24–7
Gaussian curvature 27–31, 33–7, 41–5
  of a cylinder 22, 32
  definition 30
  intrinsic measurements 30–2
  and Riemann curvature tensor 35, 77, 192
  of Robertson-Walker model 148
  of Schwarzschild space–time 42–3, 197
  of a sphere 28
Gauss's 'Excellent' theorem 33–5, 77, 192
generalized covariance 57, 63
geodesic 26, 31–3, 35, 41–2, 70–1
  definition 32
  and free fall 41–2
  null 67
  space-like 67
  as a stationary path 66–7
  time-like 67
geodesic deviation 41–2, 75–7
  and Riemann curvature tensor 75–7
geodesic equation 63–4, 67–71, 75, 89–96, 149, 186–8, 209, 215
gradient vector 50
grand unification 181–2
gravitational field 1–2, 84–6
  uniform 9, 67–8
gravitational energy release 114–15
gravitational force 1
  and metric connections 67–8, 86
gravitational lens 98–9
gravitational potential 17, 45, 106
  and metric components 45, 73, 86
gravitational waves 104–5, 124–43
  detectors 138–43
  energy flux 198–200
  polarization 127–31
  propagation 124–34, 198–200
  sources 134–8, 201–3
  velocity 126
Grossmann, M. 4, 40

Hawking, S. W. 116
  radiation 116–17
helium content of Universe 165–7
Higgs particles 171
homogeneity of universe 144–6
horizons 107–8, 111, 113, 115, 153–4, 169–72
Hubble's constant 99, 146, 151, 158–9
Hubble's law 99, 146
Hulse and Taylor's discovery of PSR 1913+16 100
hypersphere 148–9, 154
hypersurfaces
  in Robertson-Walker model 148–9

impact parameter 93
indices
  greek 39
  latin 39
inertial forces
  and metric connections 68–70
inertial frame 3, 6, 11, 149
inertial mass 10, 12–15
inflation 169–72, 183
initial singularity 147, 152, 160, 175
intrinsic curvature 22
isotropic coordinates 204

Kaluza-Klein theory 181–3
Kerr black hole 115–16
Kronecker delta 47

LeMaitre, G. 156
light cone 7–8
  and black holes 111–13
light deflection 1, 18–20, 93–5, 98–9, 204–6
light-like interval 8
local vectors 31–2, 47
Lorentz transformations 6, 46
luminosity 133, 136–7, 203

Mach's principle 3, 150
mass
  and curvature of space-time 2, 77–81
  inertial and gravitational 10, 12–15
  limit for neutron stars 118–121
  PSR 1913+16 104
matter dominated era 156–7, 160–1, 164–5
Maxwell's equations 2, 182–3
measurement of curvature 27–31, 77
Mercury, precession of orbit of 1, 87–93
metric components, *see* metric tensor
metric connections 59–62
  in classical/Newtonian limit 67–71
  conditions under which they vanish 70
  definition 59–60
  fundamental formulae 61–2, 194
  and gravitational field 68
  for polar coordinates 70–1
  in Schwarzschild space–time 195–7
  on a spherical surface 192
metric equation, definition 24
metric tensor 39–41
  associated tensor 56
  in classical/Newtonian limit 84–6
  for Euclidean spaces 40
  general transformations of 53
  for Minkowski space 38–9
  properties of 48, 55–6
  in Robertson-Walker model 147–50
  in Schwarzschild space–time 44, 196
  signature of 38
  symmetry of 55
Michelsoon interferometer 125, 138–41
microwave background, *see* CBR
Minkowski space 38–40, 65–6, 77, 111
missing mass, *see* dark matter
momentum
  conservation of 17, 80–1
  four-vector 6–7
Mossbauer effect 17–8

naked singularity 115
neutron decay 166–7
neutron star 100, 104, 120–1
Newtonian limit of GR 84–6
Newton's first law of motion 41
Newton's law of gravitaton 1, 78, 85
Newton's second law of motion 63
nucleosynthesis
  of heavy elements 165
  primordial 165–7

Olber's paradox 168–9
Oppenheimer-Volkov equation 121
orbits in Schwarzschild metric
  for light 93–5
  for material particles 88–93
  near a black hole 113–15

pair production 116, 165, 179
parallel transport 31–3, 64
  definition 58–9
  on a sphere 31–3
Penrose, R. 115–16
Penzias and Wilson's discovery of CBR 144
Planck era 173–5
Pound and Rebka experiment 17–18
precession
  of periastron of PSR 1913+16 2, 100–4
  of prihelion of Mercury 1, 2, 87–93
pressure
  degeneracy 118–21
  and gravitational collapse 79, 118–21
  in a perfect fluid 121, 154
  in universe 154–7
principle of generalized covariance 57, 63
proper time 6, 44, 62, 108
pseudo-Euclidean space 26, 38
pseudo-Riemann space 26, 40
PSR 1913+16
  analysis of motion 100–5
  gravitational radiation from 104–5, 134–6
pulsars 100, 121–3
Pythagoras' theorem 22

quadrupole moments 133–4, 136–7, 202–3
quadrupole polarization 128–32
quantum electrodynamics 176–9
quantum gauge theories 176–85
quantum gravity 173–85
quasars 98–9, 165

radiation dominated ear 163–5
recombination era 164–5, 167
red shift
  cosmological 98–9, 146, 151–2, 163–5
  gravitational 15–18, 42, 104, 109–10
resonant bar detector 124, 141–3
Ricci scalar
  definition of 82
  in Robertson-Walker model 155
  properties of 190–1
Ricci tensor
  definition of 82
  properties of 190–1
  in Robertson-Walker model 155
  in Schwarzschild space–time 197
Riemann spaces 24–7, 40, 65

Riemann tensor 73–7, 82
definition of 75
and Gaussian curvature 35
and geodesic deviation 75–7
properties of 189–93
in Robertson-Walker model 155
in Schwarzschild space–time 197
on a spherical surface 192
and tidal forces 85–6
Robertson-Walker model 147–54
rotating black hole, *see* Kerr black hole
rotating frame 9, 69–70

scalar product 48
Schwarzschild space-time 38, 42–5, 87–105, 106–15, 195–7
curvature tensors in 195–7
derivation of metric for 195–7
metric in 44
orbits in 88–97, 107–9, 113–15
simplistic introduction to 42–5
Schwarzschild radius 106–7
space telescope 146
space–time
Minkowski, *see* Minkowski space
Schwarzschild, *see* Schwarzschild space–time
special relativity 6–8, 38–40
postulates of 6
validity in free fall 11–12, 46, 59, 63, 65, 73, 111, 149
spectral shift, *see* red shift
speed of light 6, 112–13
spherical surfaces 21–4
curvature tensors 191–3
Gaussian curvature of 30
geodesics 26
metric connections of 192
metric equation for 24
static universe 83
stress-energy tensor 77–81
definition of 79
of dust 78–9
of gravitational waves 132–3
of a perfect fluid 121, 154
supersymmetric strings 183–5
supernovae 118–20, 137–8
SU(2), SU(3) symmetries 180

tangent vector 50
temperature
of black hole 117
of CBR 3, 145, 152
of universe 164, 166–7
tensor analysis 46–56
tensors
associated 55
contraction of 54
rank of 52–3
tranformations of 47, 50, 53
tidal force
near a planet/star 11
and Riemann tensor 86
and space–time curvature 41–2, 73
transformations
general 47–9
in special relativity 6–8
transverse traceless gauge 126–7

uncertainty principle 116–18, 173
unification of forces 180–3
universe
age of 146–7
closed 161
curvature parameter of 148
density of 158–9
early 163–7
expansion of 146, 151–4
inflation of 169–72
open 161
scale factor 150, 162
U(1) symmetry 177

vacuum fluctuations 83, 116, 173
vector components 47–9
Venus
precession of orbit of 93
radar echoes from 95–7
Virgo cluster 138, 144

white hole 113